ARMS RACE AND PEACE

ARMS RACE AND PEACE

By

Dr. M. Lakshmi Narasaiah
M.A., Ph.D.
Professor & Head
Department of Economics
Sri Krishnadevaraya University Post-graduate Centre
Kurnool–518 002
Andhra Pradesh (India)

DISCOVERY PUBLISHING HOUSE
NEW DELHI

Reprinted – 2025

ISBN: 978-81-7141-763-6

Arms Race and Peace

Published by:
DISCOVERY PUBLISHING HOUSE
4383/4B, Ansari Road, Darya Ganj
New Delhi-110 002 (India)
Phone: +91-11-23279245; 23253475; 43596065
Mobile: +91 9811179893 / +91 9871656464
E-mail: discoverybooksindia@gmail.com
orderdphbooks@gmail.com
namitwasan9@gmail.com
web: www.discoverypublishinggroup.com

Printed at:
Infinity Imaging Systems
Delhi (INDIA)

Preface

Until recently the likelihood of achieving a world without nuclear weapons was very, very small. But we are now living in a different world, and in this new configuration, what was a utopian dream yesterday can be the subject of serious discussions today, and put into practice tomorrow.

It was by a quirk of history that the conception of the atom bomb coincided with the start of the Second World War. The main motivation for the scientists who initiated the work on the atom bomb was that the bomb should not be used. Our argument was that we needed the bomb in order to deter Hitler from using his bomb against us. But as it happened our bombs were used, they were used as soon as they were made, and they were used against civilian populations. The bombs on Hiroshima and Nagasaki have brought the Second World War to a rapid end. But they also had another effect, namely, they demonstrated to the Soviet Union the newly-acquired, overwhelming power of the United States. From the very beginning nuclear weapons were seen as a major tool in the ideological struggle between the United States and the Soviet Union.

With the end of the Cold War, and the collapse of one of the combatants in the world power struggle, a unique opportunity was created for a radical solution to the nuclear-weapon issue. But instead, the nuclear arsenals are being maintained, albeit at reduced levels. The main reason given for this is that nuclear weapons are needed as a safeguard against the potential threat from new nations acquiring these weapons.

Horizontal proliferation is a real danger, but the retention of nuclear weapons as a means of dealing with it is about the worst possible answer. At the heart of horizontal proliferation is the perception that nuclear weapons confer power, prestige and protection. This motivated the earliest proliferators, France and Britain, and is sustained by the fact that the only five permanent members of the Security Council, with right of veto, are the five nuclear weapon states. As long as this nuclear cult exists, as long as the belief is sustained that nuclear weapons bestow status, strength and security, the pressure to join the club will be irresistible.

The main instrument to prevent nuclear proliferation is the Non-Proliferation Treaty. By now 162 states have signed the Treaty, including all five nuclear weapon states. But the NPT is an interim arrangement, a step towards nuclear disarmament, as clearly stated in its Preamble. A stable world order must be based on the rule of law, and one cannot imagine an international law that permanently discriminates between nations. If some states are allowed to keep nuclear weapons, because – they claim- they are needed for their security, one cannot deny the acquisition of these weapons to other states.

Dr. M. Lakshmi Narasaiah

Contents

1

A Nuclear Weapon-Free World–That Dream Must Become Reality

Until recently the likelihood of achieving a world without nuclear weapons was very, very small. But we are now living in a different world, and in this new configuration, what was a utopian dream yesterday can be the subject of serious discussions today, and put into practice tomorrow.

It was by a quirk of history that the conception of the atom bomb coincided with the start of the Second World War. The main motivation for the scientists who initiated the work on the atom bomb was that the bomb should not be used. Our argument was that we needed the bomb in order to deter Hitler from using his bomb against us. But as it happened our bombs were used, they were used as soon as they were made, and they were used against civilian populations. The bombs on Hiroshima and Nagasaki have brought the Second World War to a rapid end. But they also had another effect, namely, they demonstrated to the Soviet Union the newly-acquired, overwhelming power of the United States. From the very beginning nuclear weapons were seen as a major tool in the ideological struggle between the United States and the Soviet Union.

With the end of the Cold War, and the collapse of one of the combatants in the world power struggle, a unique opportunity was created for a radical solution to the nuclear-weapon issue. But instead, the nuclear arsenals are being maintained, albeit at reduced levels. The main reason given for this is that nuclear weapons are needed as a safeguard against the potential threat from new nations acquiring these weapons.

Horizontal proliferation is a real danger, but the retention of nuclear weapons as a means of dealing with it is about the worst possible answer. At the heart of horizontal proliferation is the perception that nuclear weapons confer power, prestige and protection. This motivated the earliest proliferators, France and Britain, and is sustained by the fact that the only five permanent members of the Security Council, with right of veto, are the five nuclear weapon states. As long as this nuclear cult exists, as long as the belief is sustained that nuclear weapons bestow status, strength and security, the pressure to join the club will be irresistible.

The main instrument to prevent nuclear proliferation is the Non-Proliferation Treaty. By now 162 states have signed the Treaty, including all five nuclear weapon states. But the NPT is an interim arrangement, a step towards nuclear disarmament, as clearly stated in its Preamble. A stable world order must be based on the rule of law, and one cannot imagine an international law that permanently discriminates between nations. If some states are allowed to keep nuclear weapons, because – they claim- they are needed for their security, one cannot deny the acquisition of these weapons to other states.

In the long term, there are only two alternatives: allow the possession of nuclear weapons to all states by eliminating these weapons. There can be no doubt that the former would lead to a highly dangerous, unstable world. The creation of a nuclear weapon-free world is therefore essential for peace and stability.

A nuclear weapon-free world is also called for on moral grounds. The whole fabric of civilized society is based on moral values, and if these are violated in one important area, how can they be defended in others? Security achieved by the threat of wholesale destruction, possibly genocide, is bound in the long term to erode the ethical basis of civilization.

Several arguments have been advanced against the idea of a world without nuclear weapon. One is that the genie is out of the bottle and cannot be put back. Nuclear weapons can, of course, not be disinvented, but this does not mean that we have to keep them in perpetuity. It is a hallmark of a civilized society that it can control—by national legislation or international treaties—the undesirable products of science and technology.

It is on these grounds that biological weapons have been banned, and a similar ban on chemical weapons has now been agreed to; the Chemical Weapons Convention has been signed by 156 States and comes into force in 1995.

Another argument is that nuclear weapons have kept the peace in Europe since 1945. This is a supposition without proof, but has gained credence only by constant repetition. It ignores 125 wars, with over 40 million deaths, in other continents; in Europe too we now have a bloody war. It also ignores the fact that during the past four decades there has been a relentless arms race that has resulted in obscenely huge nuclear arsenals. A more serious objection is that a treaty to eliminate nuclear weapons could be violated by a state concealing a clandestine nuclear cache, or by a later "break-out". Considering the enormous destructive potential of these weapons, such action might give the transgressing state vast power. However, this is not an insurmountable obstacle.

Even with the present state of technology, it is possible to design a system of verification that will greatly reduce the chances of undetected violation. This technological verification can be enhanced by "societal verification", that is by calling on the whole community, to report to an international authority any attempted violation of an international treaty. To be effective, this would require a clause in the treaty, and indeed in national legislation, to make such reporting a citizen's duty.

A recent Pugwash study of such schemes, as well as of methods of enforcing treaties in a nuclear weapon-free world, concluded that the problems of ensuring the integrity of a treaty to eliminate nuclear weapons is less difficult than is generally believed. The study has shown that more international intervention will be needed, such as control of all fissionable material and enforced legislation such as guaranteed protection of whistle-blowing. Measures like these will constitute infringements on national sovereignty, but limitations of sovereignty will have to be accepted in any case. We live in an ever more interdependent world, and the time has come for an extension of the loyalty to one's nation to a new loyalty, a loyalty to mankind.

Mikhail Gorbachev, whose adoption of a new way of thinking has transformed the world, was the first contemporary world leader to realize that a nuclear weapon-free world is an integral part of stable peace. He suggested the year 2000 as a target date, but this has to be understood to mean the time for agreement on a treaty, rather than for the actual destruction of the weapons, that will take many years.

The main task for the remaining years of this century is to convince world leaders, and the general public, of the necessity of a treaty to eliminate nuclear weapons, binding all nations. During this period we should also seek the implementation of intermediate steps, such as a comprehensive test-ban; adoption of the no-first-use policy; tightening of the safeguard of the International Atomic Energy Agency; and strengthening the peace-keeping and peace-enforcing powers of the United Nations. An accelerated programme of dismantlement of nuclear warheads, and further deep reductions of nuclear arsenals are of course essential steps.

The very first resolution of the UN General Assembly called for the elimination of atomic weapons. At long last, the UN is in a position to fulfil the functions for which it has been set up, and the time has come to implement its first resolution; the time has come for a decision to create a nuclear weapon-free world.

2

Nuclear Arms Race on the Subcontinent

For peace-lovers it was a strange and bewildering sight: people dancing and cheering in the streets of New Delhi and Islamabad because their governments had exploded an atomic bomb, politicians bragging about the nuclear capabilities of their countries, the press going wild over the achievements of their scientific institutions. Achievements? In Europe, people had lived for decades under the threat of a nuclear holocaust, they had danced and cheered when the Cold War ended. Now another threat of nuclear war, this time between two of the most populous, but also the most impoverished nations of the world? How could anybody be happy about the news of the successful nuclear tests? Wasn't it abusurd that the masses were cheering when their governments were spending the little money they had on the military and on expensive atomic gadgets instead of combatting poverty in their countries?

The political reactions were quick to follow. A day after India had exploded its first bombs under the Rajasthan desert, in a concerted action, the EU and the G7 countries as well as the World Bank suspended all new loans for the country. Pakistan met the same fate after it conducted its own atomic test series a few days later. Development cooperation with the two countries has thus been dealt a severe blow. In the last 50 years, both states have been important recipients of world aid.

The worldwide protests and the suspension of development cooperation with India and Pakistan because of the atomic tests may be interpreted as another example of Western hypocrisy. People remember that not long ago, the French government was the target of world-wide indignation because of their series of underground nuclear tests in New Caledonia. Stubbornly, the French President at the time rejected all criticism with the argument that the tests were necessary for the security of France and that tests would end as soon as enough scientific data had been collected. The Chinese similarly displayed total indifference to world opinion when they followed with their own nuclear test series. Not without justice, the Indian and Pakistani governments point out that they have not signed the Nuclear Non-Proliferation Treaty and the Test Ban Treaty and are, therefore, not bound by any international commitment to observe nuclear abstention. India and Pakistan can also rightly point out that the nuclear powers have not fulfilled their own commitments for nuclear disarmament which are part of the Non-Proliferation Treaty. It is really difficult to explain why the five members of the exclusive nuclear club should have the sole right to nuclear arms including the right to develop ever more sophisticated and deadly nuclear weapons. Nobody can be surprised that big countries like India are demanding equality with the nuclear 'haves', and that a hostile neighbour like Pakistan which sees itself as the rival of India on the subcontinent is trying to stay in the race by building its own bomb.

This is the political side of the coin. The quest of governments to 'keep up with the Jonesses', their claim not to be a second class power. But there is also the other side of the coin: Is nuclear equality really in the interest of the people (even if they are dancing in the street when the bomb goes off)? What do they gain from it—for their daily lives, their health, their education, the future of their children? Do their lives not become more insecure, threatened by nuclear war or nuclear accidents? The people in India and Pakistan must answer these questions for themselves. At the moment, gripped by chauvinistic excitement, a majority seems to believe that national aggrandisement is more important for them than better schools, hospitals, houses or roads.

If Indian and Pakistani scientists are able to build nuclear bombs, are they not also able to solve other technical problems in their countries? If their governments have the money for nuclear armaments, should they not be expected to also pay for infrastructure projects, for the provision of fresh water and electricity, for the running of health and education services? How can foreign donors be asked to deal with poverty reduction in the slums or rural areas of their countries, while the military claim a rising share of the national wealth for their ambitious armament programmes?

Certainly, India and Pakistan are sovereign nations and as such they have the same right as the established nuclear powers to spend their money on the atomic bomb. But if they choose that option, they should not ask other countries to help them in their social and economic development. Development cooperation already suffers from a waning acceptance among the population of donor countries. The Indian and Pakistani bombs have given "development fatigue" another push. People should know this when they rejoice over the arrival of their countries in the nuclear club.

3

The Fabric of Peace

In a world pervaded with violence, the struggle for peace must begin in everyday life. From time immemorial, peace—in the sense of peace between nations and peace within societies—has been exclusively based on the interplay between justice and force, a relationship which is at once conflictual and consensual.

History shows that peace has been and continues to be primarily an affair of state and of states, based on the use of force—in other words, ultimately, on recourse to war. This force is legitimized by highly diverse and in some cases contradictory concepts of justice.

In our time, however, the nature of war has changed. Most often it is no longer waged between states but within their borders. Conflicts within states have become so widespread that warfare has never been so rife—and in many cases so unnoticed—as it is today. Furthermore, violence exists in all our societies in one form or another, even if it does not necessarily erupt into armed confrontation. It takes many forms, and may have even been considered the norm. Its most glaring symptom is the growth of inequalities and the social exclusion to which this gives rise. Societies torn to the point of disintegration by civil war or violence are societies whose regulatory mechanisms—the bodies which exist to settle conflicts—are wrapped or paralyzed.

Some might say that this gloomy analysis heralds a future in which war and violence will inevitably prevail. The fact is, however, that these three developments—the changing nature of war, the proliferation of different forms of violence and the weakening of mediatory mechanisms, a process accelerated by globalization and the information revolution—create a new arena in which a culture of peace can emerge. And the prime mover in this culture of peace will no longer be the state but the individual, in other words each and every one of us.

For surely the road to peace must start within ourselves—in valves, behaviour and attitudes which can foster a sense of community that is today threatened. Where else can the foundations of peace be built but in our daily lives, through willingness to listen and talk things through with others on equal terms within the framework of a caring society?

The only obstacles to this enterprise are those we create ourselves, because of ignorance or fanaticism or because of the selfishness that today we are all too often asked to regard as the hallmark of human identity. This approach requires more than good intentions or the occasional act of generosity. The capacity to talk and listen to others and be receptive to their needs can pave the way to peace through an acceptance of a shared responsibility towards other people as well as towards ourselves. The mainspring of the culture of peace is making common cause with others in peace-building projects in everyday life, in whatever area of society we may be involved.

A Participatory Process

This message is not new. The culture of peace is a fabric that has been woven for generations in all societies, though its practices are not necessarily dominated by that specific title. In some places it may be known as tolerance, non-violence or justice. In others, as harmony, solidarity or conviviality. All over the world it has its defenders, some working in obscurity, others in the spotlight of public life. But its scope would be much smaller today had it not been given expression in the disinterested acts of thousands of anonymous men and women capable of listening to others, talking to them and acting with them and on their behalf.

The concept of "a culture of peace" has clearly not appeared from nowhere. But to have a single expression to describe a multigrade of ethical and practical initiatives may help to highlight their common purpose, make them more widely known and bring them together. It may sharpen the impact and focus of movements that are active in a vast range of fields and settings.

The importance of the culture of peace has now been recognized by the world community. The General Assembly of the United Nations unanimously proclaimed the year 2000 as the International Year for the Culture of Peace. In taking this step, UN Member States accept their own limitations and their urgent need for concept of peace that will be a participatory process to which all members of society can contribute no matter how humble their circumstances.

The culture of peace is intended to be a rallying point that transcends the treaties and agreements that have so often been given short shrift by history. It will become a real and living culture if we take its tenets to heart and shape a common future in our words and deeds.

Peace based exclusively upon the political and economic arrangements of governments would not be a peace which could secure the unanimous, lasting and sincere support of the peoples of the world, and that the peace must therefore be founded, if it is not to fail, upon the intellectual and moral solidarity of mankind.

4

A Universal Responsibility

Each of us is responsible for replacing the logic of force with the logic of reason and respect for the view of others. On the threshold of a new millennium, the issue of responsibility is taking on a new dimension. Humankind is still beset by war and violence. It also faces new global challenges. The impact of human activity on our planet is so great that for the first time in recorded history, we may be approaching the point of no return.

The conflicts that have arisen since the end of the Cold War erupted not as a consequence of new freedoms but in reaction to earlier oppression or repression. Suspicion, intolerance and hatred built up over decades, even centuries. But alongside the armed strife of recent years, humankind has begun to demonstrate a new skill in resolving conflicts. Mozambique, El Salvador, the Philippines, the changes in South Africa that would have been inconceivable just a few years ago, the efforts for peace in the Middle East and finally, the beginnings of a settlement in Northern Ireland are all examples proving that conflict is not inevitable. They demonstrate that breakthroughs to peace can be made by dialogue, mediation, negotiation and imagination not force.

That is why the transition from a culture of war to a culture of peace is the foremost challenge as the twentieth century draws to a close. To succeed we—all of us day in and day—out must not only do away with approaches based on force and imposition, but profoundly change cultural attitudes and daily behaviour.

We must use imagination and resolution to go to the roots of world problems and nip conflicts in the bud or, better still, prevent them. Learning to live together means daring to share, daring to do things differently, and daring to dream of a better, safer, more just and human world. It also means having the resolve and courage to transform our dreams into reality.

Here I wish to underline the pivotal role of education in promoting a culture of peace. By education, I mean not only formal instruction in schools but also informal training within a whole range of cultural institutions, including in the very first place the family and the media.

Who will be responsible for changing the culture of war into a culture of peace? Governments, parliaments, intergovernmental organizations, we might reply. The answer is correct, but it is not the only one. The transformation cannot be achieved without the active involvement of those with financial resources and influence. This answer would also be true, though only partly so. For in the final analysis, replacing the logic of force and confrontation with the logic of reason and respect for the views of others is a responsibility that belongs to all nations and all citizens, to each of us no matter how great or small the scope of our individual responsibility. The challenge of promoting a culture of peace is so broad and far-reaching that it can only be accomplished if it becomes a priority for the entire United Nations system. The dream of achieving a world without strife and violence is urgent.

Mahatma Gandhi said, "In the midst of darkness, light prevails." It is that light which is spread by the democratic values enshrined in our constitution: justice, freedom, equality and solidarity.

5

Myths and Illusions

The tide of precarity is rising steadily, so that people who have never been poor no longer regard poverty as a distant prospect but as one so close that it could engulf them at any moment.

In 1989, the fall of the Berlin Wall was rightly welcomed because it marked the collapse of a system that provided a degree of equality but rejected freedom. Today there is a strong possibility that the system gradually spreading all over the world—a kind of neo-liberal fundamentalism—may also collapse. In its obsession with freedom, vital though freedom is, this fundamentalism disregards equality, a term which should not be regarded here in purely static and statistical terms, but as something dynamic and ethical. Equality can only be truly practised in a context of social solidarity or to borrow from the vocabulary of the French Revolution of fraternity.

On the one hand, we have a world that is immensely rich in resources, possibilities, knowledge and experience; its constituent societies are freer and more dynamic than ever. There is an extraordinary potential for everyone to live a better life. But at the same time, new and ever higher walls are being built both between peoples and between social groups within individual countries. We are experiencing a travesty of development, which is creating a world bipolarized into extremes of wealth and poverty.

The most common reactions to this disastrous situation are very often the result of two misapprehensions. The first can only be described as ideological or doctrinaire since it is not based on the facts as they can be observed. It says that since the dominant system of values and things is by definition more than satisfactory, the persistence of impoverishment is merely a temporary blip. Enough time has elapsed, however, for us to see that this is not the case, including in countries where this system has been part of the established order for more than a century. One statistic is particularly eloquent. In just over 30 years, world production has approximately doubled, but the gap has more than doubled between the income of the 20 per cent of world's people living in the richest countries and the income of the world's poorest 20 per cent, according to the United Nations Development Programme.

The second misapprehension stems from another form of blindness and illusion, namely, the belief that poverty can be regarded exclusively as a moral issues as if it had no other kind of implications for those who are not poor. Globalization is, however, a two-way process. It enables the countries of the North to export their values and their paradigms as well as their goods and capital to the countries of the South, but it also makes them much more vulnerable to the backlash of crises that afflict these countries. Even in the North, the cult of competitiveness is undermining situations once considered extremely stable. The tide of precarity is rising steadily, so that people who have never been poor no longer regard poverty as a distant prospect but as one so close that it could engulf them at any moment.

Because of inadequate socio-economic development, the extraordinary upsurge of democracy over the past 30 years remains a very fragile process, and there is a risk that the trend may be reversed. When hunger, disease and ignorance prevail, citizens' participation in decision-making becomes either non-existent or a mere charade. Democratic institutions become empty shells, representational bodies existing in form only and devoid of real significance.

Social divisions caused by economic distortions exacerbate the failures of democracy which in turn pose serious threats to civil order within countries and to peace between nations. It is high time to face these obvious facts.

6

Peace and Poverty

Peace should not be understood in military terms, like absence of armed conflicts. Peace should be understood in a human way in a broad social, political and economic way. Peace should mean social justice between nations and within nations. It should mean establishment of human rights for all people.

In the new context the concept of "peace" would be the existence of a political and economic environment where each individual human being is truly free; free from the control of any powerful person or any powerful nation, free from poverty, hunger and indiginities, each individual human being free to explore the limits of one's own potential.

Today peace is threatened, more than anything else, by poverty, unjust social and economic order, absence of democracy and environmental degradation.

The cold war cloud has gone. You can feel the breath of fresh air around the world. Now there is no visible competitor left for capitalism. It is quite risky to live with a philosophy which has no challenger. To be safe, we must go to the essence of the philosophy of capitalism rather than be satisfied with the practices which emerged over years through patchworks of expediency.

Contrary to common belief, it is not the "free enterprise" which is the essence of capitalism. It is the freedom of individual thought and freedom of individual action which is the essence of capitalism. It is these freedoms which support free enterprise, free trade, free circulation of capital, and free circulation of people.

We must work out a new system, appropriate for the new world, from the basics of capitalism, not from the practices of capitalism. Many of these practices take away freedom, rather than guarantee it. Traps must go. People cannot remain trapped in places where they cannot live because of ecological, political, or economic reasons. This planet belongs to all people. If some people are trapped somewhere, we must all come forward to remove the causes of their discomfort. At the same time we must leave our shores open for anybody who decides to join us, or anybody who decides to part our company.

Poverty denies a person control over his destiny. Poverty means not being able to tell what tomorrow would be like. If we examine the situation carefully we'll see that the poverty is neither created by the poor, nor sustained by the poor. It is the system of policies and institutions that we have built around us that creates and sustains poverty. Poverty is the denial of human rights. Over one billion people live below the absolute poverty line right now on this planet, are denied of almost all human rights. There is no way one can defend the existence of poverty anywhere. Poverty is a disgrace for the entire mankind. Because we allow another human being to die of hunger, or malnutrition, or common curable diseases, or exposure to climate, we are reduced to less human beings. If a particular world system is responsible for creating this massive poverty we must act to replace it.

Resource-wise or technology-wise, there is no reason why poverty should exist and continue to deepen and widen. If we make up our minds to wipe out poverty from the surface of the earth, the worst aspect of poverty can be removed within the next couple of decades.

We can build a poverty-free world at a fraction of the cost of what we spend on war preparations. Nations become very generous when it comes to making their war-machine heftier in the name of ensuring "peace". Can we persuade ourselves to allocate a part of our time, money and intellect to achieve peace by making the people at the bottom the winners, rather than nations winning wars? "Peace" achieved by winning wars is earned by destroying people. The real peace can be achieved by building people, by reinforcing people, by helping people to reach their potential. Removing poverty is the process of building people.

Each human being is a wonderful creation of the Creator. Each human being is born with great potentials. Poverty denies any opportunity for a person to achieve any of his/her potential. We have built a world system which is in the habit of pushing people down, not building them up. It creates barriers around individuals, rather than remove them.

The most effective step that we must take to remove poverty is to create a system which create enabling conditions for people and removes the existing barriers. The institutional barriers were skillfully crafted over the centuries to benefit a handful of people.

Resource-poor nations with high incidence of poverty waste away enormous human capability each day by denying poor people the use of their energy and ingenuity. If they could have been made economically active, not only they could have contributed in the national production, they would have helped expand the domestic market for the products produced. The poor can be transformed into the engine of growth if we only allow them to unleash their capacity.

We cannot be at peace with ourselves if we know there is a human being who lives a life worse than an animal. A human being is supposed to live differently than an animal. He/she is supposed to live a life with human dignity. Human dignity is what distinguishes a human being from an animal. When we cannot ensure this dignity for others, our own dignity becomes an empty pretense.

There must be a thousand and one ways to remove poverty from the earth. We may or may not know some of those ways already. Obviously there are many more ways yet to be designed, each more effectively than others. When we shall find them, how many of them we shall find, how quickly we find them, will depend on how eager we are to find them. But to say that poverty cannot be overcome, directly and quickly, is to underestimate the capacity of human mind.

Poverty is homogeneous only when considered from the point of view of income or consumption: the uniformity of the poor as a category exists only on the level of the fact that they have little to consume. When considered from the point of view of production, i.e., the circumstances in which the poor must operate to gain their income, the conditions of poverty are extraordinarily diverse. A concrete grasp of these diverse circumstances is the first step in developing relevant instruments to address not only the problems of the poor, but also the challenge of taking advantage of the opportunities available to them.

The conventional means of measuring economic progress, such as Gross National Product per capita, tell us little about the real nature of poverty. In recent years this sort of yardstick has been supplemented by measurements of food security, income distribution, and social development (encompassing health and education). These offer the possibility of composite indices, allowing the development of more rounded characterizations and comparisons of poverty at the national level. However, these principally refer to the symptoms of poverty, not to the relational factors generating it. Poverty is not a state of being, it is the effect of dynamic processes. While it is important to know where poverty is greatest, it is critical to know why it exists. This inquiry necessarily leads away from the nature of the poor as individuals to the nature of their social and physical environment. Poverty is not only a personal phenomenon, it is a social status. As such, while its effects can be measured on the level of the individual, its causes must be sought elsewhere. From the point of view of poverty alleviation of process of becoming is just as important as the state of being.

At the heart of poverty is the inadequate access of the poor to productive resources. Low incomes tend to reflect inadequate means of production, not incompetent producers. However, poverty in India is not simply a reflection of private resources. A broad range of "external" factors impinge on incomes, among them the following:

National Policies

One of the ironies of Indian development is that while no government wants poverty, many policies contribute to it – what is given in anti-poverty programmes is drained away by other policies. The poor do not always come out ahead in the balance—they are often net "donors" to the rest of society. Frequent reference is made to unsustainable forms of development – to urban over-expansion, industrialization based on subsidies, and to public sector engorgement. What is less frequently realized is that the bill for these phenomena is often presented to the rural poor. Taxation of exports to sustain sectors with little export potential of their own and subsidized food imports to supply the urban population are policies that are often paid for by the rural poor. In many areas of India, exports are agricultural goods produced by small farmers. Here export taxes contribute to rural poverty. The same is true of "cheap" food imports which depress the prices paid to small farmers for their food crops.

"Structural imbalance" is not only a recipe for increasing external indebtedness, it is also a recipe for increasing the poverty of the rural population. The political weakness of the poor in most areas is not only the basis for inadequate poverty alleviation programmes and policies—it is the basis for an actual transfer of their income to more socially influential groups. While it is often correctly asserted that the poor are the first to suffer from adjustments involving public social expenditure cuts, it is often the case that they also have the most to gain from the elimination of policy-based economic distortions that reflect social power rather than productive efficiency and potential.

Demographic Factors

Accelerated population growth is a long-term contributor to poverty. In India the incomes of the poor have declined, mortality rates are also falling, pushing the numbers up. In the meantime, land is becoming scarcer, plots more fragmented and the soil and pasture increasingly degraded. This phenomenon is not without its policy dimensions. As long as the poor remain undercapitalized, and essential determinant of household income is the amount of labour available to it household economic strategies favour large families. While population policy has a role to play, possibly more critical is a change in the economic environment. Access to capital and more secure income changes perceptions of the need for labour. In the medium and long-term, population dynamics are driven by the underlying productive systems. As long as the production systems of the poor remain underdeveloped, population growth remains high, restricting even the future possibility of development.

Natural Resource Management and the Environment

If poverty is both cause and effect of rapid population expansion, so poverty is both cause and effect of many dimensions of degradation of the environment. Many of the rural poor, but by no means all, live in areas of extreme environmental fragility, a circumstance often prompted by high level of control by the better-off over more stable and productive resource areas. Here the poor are extraordinarily exposed to the dangers of erosion, whittling away at an already meager productive base. The threat is not entirely due to nature. Rather, poverty accelerates erosion. Without capital, the poor are frequently unable to invest in even traditional methods of soil and water conservation. And without sufficient land they are forced to shorten fallow periods, putting further strain on the resource base. As in the case of population growth, the result is strain not only on the poor, but on the entire Indian economy. Given the extremely limited economic alternatives, the solution to this problem is not to forbid the use of environmentally fragile resources to the poor, it is to change the conditions under which their use takes place. An access to conservation technology is important; but more so are security of land tenure and resources to invest.

Combating poverty means not only increasing the production of the poor, but also preserving and enhancing the long-term value of the resource they control. What this very often means, in practice, is assisting the poor in reestablishing a stable relationship with fragile resource. Prevailing processes in many areas involve the gradual – and sometimes not so gradual— depletion of natural resources, to the detriment of all. Part of the answer to this is conservation. Part of the answer is also to provide viable economic alternatives to the poor, reducing their dependence on erosion-prone crop and livestock practices.

Exploitative Intermediates

The poor are not unaware of the pressure upon them, and also of means of overcoming them. Their ability to respond, however, is severly impaired by social powerlessness. The poor are surrounded by a dense network of public and private factors reducing their freedom of action, and actually draining what few resources they do have. Members of the network include traders and moneylenders capitalizing upon the economic weakness of the poor, and engaging them in unequal exchanges. They also include public agencies either indifferent to the requirements of the socially uninfluential, or actively engaged in extracting "surplus" for use by other groups. Not to be excluded from this are organizations which are ostensibly "for" the poor, but which, in fact, serve as systems of containment and control.

7

No Progress without a Secular Society

Every day, women continue to be victims of rape, trafficking, acid-throwing, dowry deaths and other kinds of torture. At the opening of this new century, women are still not considered as equal human beings in many parts of the world. Religion and patriarchy continue to have an all-encroaching hold on their lives, maintaining and justifying their age-old oppression. In some South Asian societies, this hold is even increasing.

I do not believe that there can be real equality in a society dominated by religion. Western countries speak repeatedly about the necessity of economic development to alleviate poverty. But this is not enough. Some oil-rich countries may be economically developed, but women are deprived of all rights. The supremacy of religion is incompatible with freedom of expression, women's rights and democracy. This is why I see religion as the main enemy of women's development.

We have to act on several fronts at once. First of all, improving access to education. In a society like Bangladesh, 80 per cent of women are illiterate. For centuries women have been taught they are the slaves of men. It is very hard to change their minds, to make them aware of their oppression, to give them a sense of their independence. This educational effort has to go hand in hand with a secular feminist movement in society. Such movements have to start within the country and they cannot take hold when people are uneducated and unaware of their oppression. I'm not sure you can accomplish much from the outside, except to expose in the media the atrocities women in all too many countries face in their day-to-day lives.

In some countries, this movement is emerging, but very timidly, and it has a slim margin of maneuver. It has the uphill task of fighting for the repeal of religious laws and the introduction of a uniform civil code. So far, it tends to be constituted by a few individual feminists who are forced to be diplomatic, to compromise with fundamentalists, be they men or women. But they are trying to change the system, step by step, and it will take a very long time. People are not yet ready to do away with religious laws that impact upon every aspect of society, from education and health to the workplace and the home.

For women's status to change, we also need enlightened leaders who believe in equality. In countries of South Asia women with a strong voice do not have the support of political leaders, whether they be men or women. Look at the countries in which women are in politics, or even heads of state. Does it follow that women in those countries are emancipated? Because of long-standing vested interests, such leaders continue to back measures that oppress women. They are not ideologically committed to changing these conditions. In South Asia, most of the women who become heads of state are religious, and like men, they adhere to the religious objectives of the Establishment. Until a society is not based on religion and women are considered equal to men before the law, I do not think that politics will advance the cause of women.

Until a society is not based on religion and women are considered equal to men before the law, I do not think that politics will advance the cause of women. In Western countries, women are educated, they are treated equally, they have access to jobs. In these conditions, their participation in politics has a meaning.

Education, a secular feminist movement, and leaders—both men and women—committed to equality and justice. This is what it will take to change the dire conditions which too many women still face today. It will take a very long time, but we are here to work towards that end.

8

Who Suffers from the Asian Crisis

It's the same old story. The people who suffer most from the effects of the crisis are not those responsible for it but it is the masses of the working population who are now thrown back into abject poverty. It is true—in Bangkok, Jakarta and Seoul a few heads rolled as governments changed, banks closed and enterprises went into bankruptcy. In Indonesia, it was even the ageing dictator Suharto himself who became a victim of the crisis. But who has heard that his immense fortune was impounded or that his widely entrenched family clan has given up their huge conglomerates of banks, industrial firms, and other enterprises through which they had siphoned off the country's wealth for decades? The government says it is "very serious" about achieving economic reform and sound governance, as well as on providing urgent social safety assistance for the poor. But before any results can be achieved, the country will have to go through valley of despair.

The World Bank says in a recent report on Indonesia that no other country in recent history has suffered such dramatic reversals of fortune. The statistics are indeed staggering. In a span of a year, Indonesia's currency lost 80 per cent of its value against the dollar, and inflation soared to over 50 per cent. Twelve to 20 million people are unemployed. If one considers that each of these people has dependants who lose their main source of income estimates are realistic who speak of 60 to 100 million people who are thrown into poverty. For this year, the World Bank expects an economic contraction of 10 to 15 per cent, and a further drop of 2 per cent in 1999. It is hard to imagine what this means for a country like Indonesia which had been proud to be so extraordinarily successful in the fight against poverty: from 1970 to 1997, the number of people below the national poverty line dropped from 60 to only 15 per cent. And now this unprecedented setback threatens to undo decades of positive development. A crisis compounded by a severe "EI Nino" drought and devastating locust invasions which led to large-scale damage to crop and are causing famine in rural areas which used to be the food basket of the nation. It is a human catastrophe which is taking place in Indonesia, one that needs our urgent attention.

Although Indonesia may be the country hardest hit by the financial and economic crisis, people in other states of the region also feel its effects. In Thailand, the national currency, the bath, is worth 40 per cent less than last year, and food prices have doubled. While foreign workers who came from neighbouring countries during the boom years have been sent home, unemployment among the Thai nationals is expected to jump to 8 per cent. Life for residents in Bangkok's slums has become tougher than ever with families breaking up and children being sent into the streets to fend for their own lives. The level of violence is going up and child prostitution is often the only way for the poor to survive. The Thai government speaks of "a fragile stability" and says that "there has been substantial progress in implementing structural reforms". But for the poor people in the country—and their numbers are getting bigger every day—there is no light yet at the end of the tunnel in which they are finding themselves so unexpectedly.

Meanwhile those who brought about the sudden downfall of the "tiger economies" of Southeast Asia – the managers in the international money markets who first invested tens of billions of dollars in the region and then suddenly withdrew the funds—are biding their time for a profitable comeback. At the moment, the International Monetary Fund (IMF) and the World Bank together with other donors are pumping enormous amounts of bail-out money into the affected states in order to prevent a total collapse of the international financial system. As soon as their emergency operation shows results and growth in Southeast Asia begins to pick up again, the speculators and investors will be back to profit from any new boom. Already now, aided by the liberalisation of the Asian economies which is a condition of IMF loans, takeovers of bankrupt Asian banks and industries by foreign companies are the order of the day. By the time the Asian crisis is over, a much larger sector of the Asian economies will be in the hands of overseas transnationals.

It is the poor who are footing the bill. In Indonesia, the current daily minimum wage is 4,400 rupiah or about 30 US cents. Reports say that even this low wage is not always paid by employers. Even in a country like Malaysia, which had hoped to reach a European standard of living by the year 2020, poverty is once again on the increase. The danger is that the economic crisis will in the long run also cause social and political disintegration as it has already done in Indonesia with the outbreak of ethnic conflict between people of Malay and Chinese origin. The governments in the region would be well advised, therefore, if they paid all their attention to the social effects of the crisis—and did everything in their power to distribute its burdens as fairly as possible. The poor who had no hand in the making of the crisis should not be the only ones to suffer.

9

Revisiting Bretton Woods–Reforming the World Trade and Finance System

That leading trio of major multilateral economic institutions (the International Monetary Fund and World Bank in Washington D.C. and the WTO in Geneva) were created from the ashes of World War Two to build a strong, coordinated, international set of economic arrangements. They did well. Their contributions significantly forge systems of cooperation between governments which, in turn, encouraged global economic growth and development.

But, is it time now to revisit Bretton Woods, that location in the hills of New Hampshire, where half a century ago U.S. Treasury Secretary Harry Dexter White, British economist Lord Keynes, and many others, set the plans for the post-war multilateral economic system.

The question is not academic. It was being asked recently in an unprecedented scale in the annual meeting of the IMF and World Bank in Washington D.C. The questioning came for three critical reasons:

First, there is a widespread view that a strong supranational institution is urgently required in the currency arena. The IMF has been absorbed with medium-term economic assistance programmes and appears to be attaching low priority to its original purpose: the IMF's Articles of Agreement declare the Fund's purpose is: "To promote international monetary cooperation through a permanent institution which provides the machinery for consultation and collaboration on international monetary problems".

Second, the WTO has brought tempers to the boil in many developing countries and created fears. The WTO's failure is serving now as a stimulus for the growth of regional trade blocs, based upon major industrial countries and open to relatively few developing countries.

Third, the World Bank has taken a backseat when it has come to advancing Western support for the poorest nations. There was a time when the President of the World Bank would use his office to rally international opinion and publicly urge the industrial nations to take a more constructive and more generous approach to the developing nations. In recent times the leadership of the institution has been silent. It has become mired in administrative maters, willing to bow to IMF leadership and content to seek to influence development thinking through the publication of economic research reports.

World Bank Subordinate to IMF

At the same time, the World Bank has come to play second fiddle to the IMF. The Fund has engineered itself into a position of leadership in economic policy discussion with developing countries and with the former command economies of East Europe and Central Asia. The World Bank does not provide programme lending of any kind until a borrowing country first has an IMF programme in place. While the two institutions are totally distinct in legal and financial terms, the Bank has accepted a subordinate position to the IMF.

These three phenomena are not encouraging for the health of global economy and from the perspective, in particular, of the developing countries.

On the monetary front there is a need to protect the interests of developing and emerging countries from the vagaries of the super-economic powers. Most of the governments of the world have looked on hopelessly as Japan, Germany and the United States have pursued nationalist economic policies that have played havoc with the currency system. Most of the world's trade is booked in the currencies of these three countries and when those currencies spin out of control, so concluding trade deals and securing investment agreements becomes far more complex.

Uncertainty and instability in the world's currency systems are menaces that the IMF was expressly designed to counter. But the IMF has become so engaged in development lending (it now talks of providing programs to some 80 countries) that its need for financial resources of its own is growing rapidly. That need makes it difficult for the Fund to be critical of its most powerful members. It cannot bite the hands that feed it. Thus, calls to the major nations for fiscal restraint, monetary discipline and international cooperation, are made in muted tones.

The Fund, however, must respond to the mounting recognition that some supranational mechanisms are needed to survey the international economic landscape, to ring the alarm bells, to push and shove for meaningful consultation and to place blame on those whose policies are so nationalistic that they endanger the international system. IMF surveillance of the major economic power needs teeth.

The Fund should concentrate once again on using its influence and its expert staff to enhance international understanding of the complexities of the global trading and financial system. By this means it can rebuild its influence with the major powers. While it is unrealistic at this juncture to call for the IMF to become the world's central bank, it could serve as a vitally important covener of consultative processes designed to attain the objectives that its founders decreed: "To facilitate the expansion and balanced growth of international trade, and to contribute thereby to the promotion and maintenance of high levels of employment and real income and to the development of the productive resources of all members as primary objectives of economic policy".

The IMF's role should be enhanced. It should blend its monetary miles with new trade roles. The WTO has been the forum for negotiations and for the supervisions of agreements. WTO does not undertake projects, it does not have powers to influence the policies of its most powerful members and it does not have the prestige needed to provide real leadership. It is time that the WTO was merged into the IMF.

Trade and Finance Belong Together

It makes little sense to split issues of international capital flows from trade questions. The globalization of trade and investment has brought these disciplines close together. If forging satisfactory agreements is often difficult, then this in part is due to the fact that distinct organizations have leadership for distinct parts (WTO for trade and IMF for money) and there is no effective mechanism for cooperation. It is also the case that within national governments the trade and finance ministries are often in conflict and face insufficient pressure to coordinate. If the IMF managed both trade and monetary negotiations on the global scale, then this would add pressures on trade and finance ministers to work together.

Returning to its original monetary roles and adding a major trade role should be more than enough to keep the IMF busy. It would be logical, particularly in such circumstances, that the Fund return to the World Bank the development financing roles that it has assumed in recent years and that diverted it from its original purposes.

The IMF's Articles stress that one of its purposes is "to give confidence to members by making the general resources of the Fund temporarily available to them under adequate safeguards, thus providing them with opportunity to correct maladjustments in their balance of payments without resorting to measures destructive of national or international prosperity".

The Fund might argue that the World Bank should confine itself to infrastructure and social project finance and technical assistance and leave all programme lending to the IMF. The reality is that the World Bank discovered to an increasing degree, starting with experiences with Turkey in 1979 and then with many highly-indebted nations from 1982 onwards, that the best development projects will fail in countries where wholly unsatisfactory economic policies are in place. The Bank has also recognized the pain and complexity of adjustment and that countries embarking on adjustment policies enter upon a multi-year process: a process better geared to types of financing arrangements that the World Bank can offer, than those provided by the IMF.

Avoid Duplication of Effort between IMF and World Bank

The experiences of the last decade have strengthened the World Bank's understanding of macro-economic policy reform and enhanced its capacity to provide comprehensive policy from support to its member countries. Cooperation with the IMF has improved, but it is also second best option and an expensive one. There remains too much duplication between the Fund and the Bank. The biggest cost is paid by the borrowing countries—ministers and their immediate subordinates spend endless hours negotiating separately with IMF and World Bank teams and developing duplicative reports.

Reform is only necessary when things are not working well. Today, there is enormous scope for improvement in the global trading, monetary and development areas. The three prime institutions created for these areas are not performing well enough. Reform is urgent: the WTO should be merged with the IMF, the IMF should refocus on issues fundamental to securing a healthy global monetary (and trading) system and withdraw from the aid game; and the World Bank should have enlarged scope and provide more leadership on the development front.

Such reforms will not end the problems that our world economic system faces and their significance will be largely determined by the support they receive from the leaders of the most powerful industrial nations, irrespective of the zeal of the officials within the IMF and World Bank. Today, in the midst of prolonged international slump where nobody is satisfied with the ways in which the international system is operating, there is an important opportunity to secure backing in the capitals of the world's super-economic powers for the types of reform that are articulated here.

10

A New World Order for Whom?

"Four holocausts" humanity has produced are breeding the seeds of our own destruction: war and militarization, human oppression, economic destitution and environmental destruction. The "new world orders" on offer can satisfy only the minority of the world's rich and will ultimately only exacerbate these four trends towards global annihilation. But there is also hope in the "grassroots world order" leading to the civil society, democratization and social mobilization as the only way for the planet to survive.

As we approach the end of the 20th century and a new millennium, humanity is faced with four conditions of its own making so serious in terms of their present destruction of life and risks for the future that they warrant a description as "four holocausts".

The first holocaust is that of war and militarization. After the Gulf War it is clear that, far from preparing the way for world peace, that conflict has unleashed a new global arms race in the weapons whose brutal effectiveness was so clearly demonstrated in Iraq. The second holocaust is that of human oppression, the violent denial by governments of the basic personal, civil, political and economic rights of their citizens, which routinely persist in a majority of countries of the world.

The third holocaust is that of economic destitution, the mass poverty of a fifth of the world's population, leading to endemic malnutrition, disease and death, not least among children. And the fourth holocaust is that of environmental destruction, which is gradually or not so gradually rendering the planet uninhabitable, as witnessed by the growing millions of environmental refugees whose bankrupt ecosystems can no longer support them.

None of these circumstances is entirely new to our age, of course. War, repression, destitution and ecological degradation have often been part of the human condition. What is new about the current situation is its global nature. All humanity, the whole earth, is now at risk. And it is all humanity which must be party to a response, if it is to be successful.

Against this threatening background, several new developments stand out. First is the collapse of communism, taking out of the bounds of credibility socialism's millenarian dream of abolishing the market and inevitably replacing capitalism through the onward march of history. Second, there are the twin trends of globalization and interdependence: globalization especially of the economy, interdependence especially through environmental impacts. This is the context for any discussion of a "new world order".

The New World Orders

There are, broadly, three kinds of new world orders currently on offer. The real world is almost certainly going to be mixture of all three, but the balance between them will be crucial in deciding whether we will successfully manage to face, and overcome, the holocausts raging among us.

The first kind of new world order may be called the "neoliberal". Its most important component is the untrammeled operation of what we would call the global "free market". At once we must qualify this terminology by noting that the freedom bestowed on someone by the market is in direct proportion to the amount of property owned by that person within it. In a free market, those who own the means of production are free to produce what, when and where they want, and largely to determine the conditions of production. Those who own the means of consumption can similarly scour the world for products to satisfy their wants. The extent to which this "freedom" is far from universal is shown by the fact that, with regard to consumption some 23 per cent of the world's population control

85 per cent of the income which is consumptions prerequisite and ownership of the means of production of more concentrated still. The neoliberal world order thus represents a good deal for perhaps a quarter of the world's population, but has precious little to offer the rest.

A Global Order Framed by International Institutions

The second kind of new world order may be termed the "social democratic" and is roughly that advocated in the 1970's by the proponents of a New International Economic Order. These two new world orders are clearly different but they also share several characteristics which seem to be more significant than their differences. First, they are explicitly western-oriented and homogenizing. They view the world through the eyes of western science and western culture simultaneously devaluing the knowledge and accumulated wisdom of the great majority of humankind. The paradigm society, towards which all others are supposed to be developing or aspiring to develop, is that of the United States.

Second, both these world orders are economistic. Human progress and development to them means economic developments, still usually measured by the level and growth of GNP per person. No social or cultural tradition or aspiration is allowed to stand in the way of this "development". Third, both world views envisage top-down decision-making for administration and control. For the neoliberals the dominant influence is exercised by the owners and managers of transnational capital. For the social democrats their influence is balanced by the interventions of international and national bureaucrats. Neither world order places great store on consultation with, let alone decision-making by, ordinary people in their communities.

In contrast to these two new world orders, it is possible to posit a third, here called the grassroots new world order, which takes as its principal focus neither the market nor the state (national or international), but civil society, the networks of family, community and voluntary association acting for social reproduction, reconstruction or reform. This new world order has characteristics diametrically opposed to those shared by the first two discussed. Its impulse derives explicitly from the bottom up, drawing on the capability and creativity of those united by shared values and interests, at the local level or in wider networks. Their world-view is one of cultural diversity, of one world comprised of many different villages rather than a homogeneous global village modelled on the US. They perceive human development to be holistic, with the economic dimension integrated with, or embedded in, a broader, social, ethical and ecological reality. And they proceed from an ethical basis that strives for ecological sustainability, social justice in distribution and broad participation in cultural, political and economic life.

Looking to the Grassroots World Order

After this thumbnail sketch of these three views of different dominant global processes, one can ask which of them or, more realistically, what balance between them will be best able to put an end to the four holocausts tormenting humanity. There is only one convincing answer: the dominant thrust must be towards the grassroots new world order. There are several reasons for this. Most obviously, it is indisputable that it is the forces of the market and the state that have not only failed to douse, but have actually fanned the flames of all four holocausts. It is states that go to war and waste the commonwealth on weaponry. It is states that are responsible for the great majority of violence and repression against ordinary people. It is states, often supported by multilateral governmental organisations such as the World Bank and IMF, that have the so-called Third World intervened massively in the subsistence, largely nonmarket economies of he people, and redistributed their resources, redefining their very rights to property, in favour of industrialization and market exchange. But these enhanced markets have then spectacularly failed to provide alternative subsistence for those dispossessed, leaving them by the million impoverished, marginalized and destitute. Moreover, this process has set in train two great engines of environmental destruction: industrialization itself, with its toxic pollution, soil erosion, water and ozone depletion and climate destabilization; and the depredations of the rural dispossessed, forced into forests or onto marginal lands, deforesting, making deserts, extinguishing species and multiplying in numbers in their desperate efforts to stay alive.

Civil Society Mobilizing for Human Survival

In contrast, it is civil society that has mobilized explicitly against the four holocausts. The great social movements of our time are those for peace and human rights, for justice and development, and for environmental conservation. It is independent, non-violent associations of civil society that have sought explicitly to address these issues, meeting at best indifference from organizations of the market and the state, at worst outright hostility.

This is not at all to say that the market and state are irredeemable, or that they have no role in combating the four holocausts of destruction. On the contrary, they each have a vital contribution to make but, it order to make it, each must first be transformed. The market failures caused by great concentrations of wealth and power, ubiquitous externalities, the tyranny of small decisions and positional goods can only be resolved by determined and democratic state action. But most governments are not democratic. On the contrary most are unrepresentative and self-serving, many are vicious and corrupt. And the only force that can democratize them is that of civil mobilization and organization.

Viewing the immense power of today's concentrations of wealth, and the remoteness of most governments from their people, one may feel despair at the prospect of civil society being able to harness these forces to the common good. And indeed there is no certainty that they will be thus harnessed. The four holocausts may simply run their awful course. But since 1989, at least, there is proof positive in the peaceful revolutions in Eastern Europe and the Soviet Union, that civil society can overturn seemingly omnipotent despotic structures. And all over the world the various movements for peace, justice and the environment are organizing for human survival. The stakes have never been so high and the outcome is uncertain; but there are legitimate, and inspiring, grounds for hope.

11

Trading Towards Peace

The reason why trade has such a vital part to play in building peace is because it means lowering barriers—not only to goods and services but among nations and peoples. The elimination of barriers creates interdependence and interdependence creates solidarity. The history of the last fifty years has shown us all the undeniable benefits of lowering trade barriers and opening economies.

Clearly every region has its own characteristics, and it would be wrong to imagine that the same blueprint can apply everywhere and in the same way. Any region which was for thousands of years at the crossroads of world trade should regain its place in the centre, because doing so will help build peace as well as prosperity. This is why the numerous applications for accession to the WTO from various countries are so significant. The first is through regionalism. There are several efforts at regional trade and economic initiatives among countries, and that such initiatives will be encouraged to reduce positive results. Regional initiatives are important because they can help countries at a comparable level of development to move relatively quickly in opening their economies and in deepening their interdependence.

However, the rapid advance of global economic integration means that while regional initiatives remain important, they are not sufficient by themselves to address successfully the new perspectives of the international economy. That is why there is a need for second track, which is the rule-based multilateral system. And that is why the multilateral system is of fundamental importance to the economic prosperity of any region.

As the first major international institution to be created in the post-Cold War era, the WTO offers a promise of the kind of global economic architecture which need in the coming decades. Its culture is firmly rooted in the tradition of consensus-building and cooperation among sovereign countries. And the WTO embodies rights and obligations negotiated by consensus, approved and ratified by each government and each Parliament, and they are enforceable, not through the crude exercise of economic power, but through the rule of law. The alternative would be a power-based system—who would want to choose this option?

But most importantly, the WTO is an organization which brings all countries—from all corners of the world and from all levels of development—together as equals. There is no weighted voting, no exclusive clubs, no inner and outer circles. Developing countries representing 80% of constituency sit as equals with industrialized countries to write the rules of a shared trading system.

This new unity of developing and developed countries inside a single system will be credited as the greatest achievement of the multilateral system. But this unity is still fragile: we cannot allow it to be broken: This is why, in preparing the agenda of the first Ministerial meeting in Singapore, have recognized the particularly difficult task facing developing countries in implementing the Uruguay Round commitments. They have also acknowledged the challenges they face in contemplating the necessary work programme.

The integration of developing countries as equal partners in the multilateral system is one of the most important challenges in shaping the economic order of the 21st century. This is a shared responsibility of developed and developing countries alike. There is no rational alternative to this objective. The evolution of the global economy makes that clear.

Now there is a need to work together as equal partners to ensure the full integration, and all other developing and transition economies, into the global economy and the rule-based multilateral trading system. In conjunction with this there is a need to encourage, notably the growth of regional economic cooperation. The alternative is a vicious circle where economic isolation feeds greater political instability which in turn leads to great economic isolation. The road to a lasting peace in the world begins, not ends, with economic integration and interdependence. Taking this message to heart will help build a future where it is goods, services, and investment that cross borders—not missiles and soldiers.

12

The End of the Old Order–No Guarantee for Peace and Prosperity

In the last few years, the world has witnessed the collapse of communism and the end of many authoritarian regimes around the globe. In Eastern Europe and the former Soviet Union, multiparty systems with free elections have been introduced. In Afghanistan and Cambodia, Ethiopia and Angola, Nicaragua and Peru leftist governments of various shades have given way to more pluralistic forms of government or are in the process of doing so. In South Africa, the white minority has accepted the black majority rule. All over the world, the old order which was established after World War II is breaking down. Freedom, democracy, self-determination and economic prosperity are the slogans of the revolutions which have swept the repressive regimes away.

But after the initial euphoria over the unexpected successes of the democracy movements, the world is now waking up to the sobering recognition that freedom does not automatically lead to peace and economic progress. In fact, in many countries which have managed to throw off the communist yoke ethnic rivalries have been sharpened by the call for self-determination. Age-old historical animosities between various nationalities and religious groups have come to the fore and are threatening to undo whatever gains have been achieved by introducing a pluralistic system of government.

The former Yugoslavia is a case in point. The state was an artificial creation born after the First World War and the demise of the Austrian-Hungarian and Ottoman Empires. Similar to many of the artificial states in post-colonial Africa, Yugoslavia consisted of a diverse mixture of ethnic and religious nationalities held together primarily by the charisma of former President Marshal Tito and the socialist ideology he imposed on the country. The wind of democratic change which blew across the European continent in the wake of Gorbachev's perestroyka also meant for Yugoslavia and hence it is divided between different ethnic or religious groupings.

What has happened in Yugoslavia is also taking place, although to a lesser extent, in Georgia, Armenia. Azerbeidjan, and Moldova where different nationalities have taken arms against each other to fight for their right of self-determination. In Ethiopia, the right to secession has been conceded to the Eritreans by the new democratic government, but already the Oromos and other ethnic groups are threatening to break away from the common state. In Afghanistan, the defeat of the communist regime has not brought peace to the country but the danger of a prolonged bloody civil war between the victorious guerrilla factions. Even in South Africa the end of apartheid marred by intensified inter-ethnic fighting between different African parties and groups.

The conclusion to be drawn from this review of recent developments around the world is certainly not that the call for freedom and democracy will inevitably lead to chaos. However, these developments must warn us that the wonderful concepts of Western philosophical thought will only work in practice if they are accompanied by the necessary spirit of compromise. Democracy functions well in societies which are very homogeneous like many of the Western European ones. Even there the system may fail when faced with deeprooted antagonisms as the case of Northern Ireland shows. But in countries which are divided by nationality, race, or religion freedom, democracy, and the right to self-determination may lead to even sharper conflict unless there is a genuine give-and-take between groups which grants every individual and every group the right to develop according to their own ideas. The older order is breaking down. The communist regimes have collapsed, most of the military dictatorships are on their way out. But in many parts of the world it is not yet clear what will come in their place. One thing is certain: there is no easy way to peace and economic prosperity. The new systems that emerge will have to guarantee the rule of law and the protection of human rights. And they must try to instill a spirit of common values which bridge the cleavages which separate ethnic and religious groups. The oppressive regimes which held together their populations by force must be replaced by governments which have learned the art of compromise—which is the essence of pluralism and democracy.

13

South Asia Quarrels over Water

Bangladesh, Bhutan, India and Nepal have at least three features in common; they are all situated on the southern slopes of the Himalayas, they have high levels of poverty and they are all rich in water resources. To these add a fourth: they frequently quarrel about water.

For decades, policy-makers in the four countries have grappled with the problem of how to harness the rivers that flow from the mountains and serve the region's over-one billion people. Mostly, they have gone in for large dams and grand irrigation schemes. But though these promise a lot, their usefulness is being increasingly questioned by many independent engineers, scientists and social activists. And they are thought to contribute to water conflicts between countries.

Large hydro-power projects represent a development model that is biased toward industry and urban areas. And irrigation schemes favour rich farmers with large farmlands, most useful for water—and fertilizer—intensive cash crops that poor peasants cannot afford to grow. Because of their size, they also cause huge displacements—again, of poor farmers. Social activities say that while such projects may have increased overall food production, they have also increased the rich-poor gap in a region that is home to the world's largest number of absolute poor-people living on incomes of less than one dollar a day.

In tandem, many South Asian hydro-power engineers and economists say 'national' and top-down engineering solutions to water management ought to be replaced by a new 'basin-wide' or regional approach through environmental and socially benign projects.

One example of such 'national solutions' is the Indian Farrakka Barrage on the Ganges. Built near the Bangladesh border in order to divert water to the Calcutta port, it has led to problems with falling water tables and salinity downstream in Bangladesh. And although Indian and Bangladesh resolved the long-simmering dispute in 1996, Dhaka is now proposing another expensive Ganges Barrage to solve the problems created by the first barrage.

Of the four countries, Nepal has the highest per capita potential for hydropower generation—its narrow valleys and steep terrain mean water flows faster, which is good for power generation. But although Nepal could potentially turn its rivers into 'hydrodollars', Himalayan dams are expensive to build and the country's experience in dealing with India on joint river projects has been patchy. At the same time exporting power to India, where powercuts that can last up to 12 hours at a time in some cities, means foreign exchange for Nepal.

Because many Napalese perceive past border irrigation projects as unfairly benefiting India, joint Indo-Nepal river projects are a political issue. A recent treaty to harness the Mahakali river on Nepal's western border with India is a case in point.

Although India and Nepal agree on equally sharing the 7,500 megawatts of power from the project, negotiations have been stalled over Nepal's demand that it be compensated for downstream irrigation benefits to India. Nepal is not allowed to harness water in many of its rivers for its own use without India's concurrence these rivers flow into India, and Nepal is bound by past bilateral treaties.

As one official at Nepal's Water and Energy Commission put it: "What's in it for Nepal? Why should we write off benefits from the Mahakali for future generations of Napalis? Things look even bleaker for larger joint projects like a planned 10,800 mw dam, on the Karnali river—the $ 10 billion Chisapani dam project is expected to displace some 60,000 Napali farmers. In addition,

India has already used up all the natural flow of the Karnali for its own irrigation downstream.

Things are different when it comes to Bhutan: India and the tiny Himalayan kingdom have sorted out their river projects without any major difficulty although experts say it is too early to assess how much the Bhutanese will really benefit. According to an estimate by an Indian water consultant, Bhutan can generate up to 20,000 mw of power from its rivers.

Generating funds for large projects does not seem to be a problem. Traditionally, the World Bank has been one of the major financiers for dams worldwide having undertaken 400 projects involving dams since 1970. In 1988, India alone had about 473 dams under construction, with the Bank involved in 45 of them.

When the World Bank recently pulled out of some controversial mega projects, such as the Arun 111 dam in Nepal, multinational companies stepped in. In Nepal, three medium-sized hydropower projects worth more than 300 million dollars are being built with foreign direct investment.

FDIs are backed by organizations like the Japanese-managed Global Infrastructure Fund (GIF) which is interested in putting money into such projects as the 270 metre-high Kosi High Dam in Nepal, the Ganges Barrage in Bangladesh and the mammoth 20,000 mw Dhihang dam in India. Funds also come from the Manila-based Asian Development Bank.

But many resource economists maintain that the chances of making colossal mistakes is higher with large projects. It is all a question of risk-management. Can our countries afford to take the risk of spending billions on doubtful projects that can go dangerously wrong?

14

Science to What Purpose?

Are the welfare and interest of the public being served by the priorities of researchers, the thrust of their work, the ways in which they are organized, the funding they receive, and the circulation of their findings?

Science reigns triumphant. Never has it been so powerful and influential. It has conquered diseases which have decimated whole populations. It has abolished exhausting physical labour and wearisome repetitive tasks. It has vanquished distance and pushed back the frontiers of our knowledge of the infinitely large and the infinitely small, in both the inanimate and the living world.

In short, it has acquired the ability to shape our lives, to change life itself. But it has also increased its capacity to destroy life. The strength of an army can rest on the number and determination of its combatants but it is also, and chiefly, based on the technological sophistication of their weaponry. The bombing of Iraq, and now of Serbia, are the latest examples.

Yet science is wavering. For the first time since the Enlightenment, the way science can be used is being challenged. The link between scientific progress and social progress is weakening and signs of obscurantism are appearing. Hiroshima sounded the alarm. Then the crisis of the environment, triggered by the dominant mode of development, turned questioning of science into a worldwide issue. This form of development is inseparable from a frantic and indiscriminate quest for technological innovation. Finally, advances in biotechnology, which harbour many grave dangers for human dignity, are often too closely bound up with the selfish interests of their promoters.

No one blames science for not knowing everything. No one criticizes it because it has not yet found a vaccine against AIDS or reached a conclusion about the theory of the Big Bang. It has never been claimed of science, as it has of history, that it has come to an end. It must keep to tirelessly probing the enduring mysteries of life.

But science can no longer avoid—and nor can we—the basic question of what and who it is for. In other words, are the welfare and interests of the public being served by the priorities of researchers, the thrust of their work, the ways in which they are organized, the funding they receive, and the circulation of their findings? Or are scientists looking mainly in the direction of high-spending consumers at the expense of long-term basic research? Because of the growing "privatization" of research, are we not tending to overlook essential and universal human needs which cannot immediately be met?

Those who are excluded from this new "scientific power" must make their voices heard. For example, the inhabitants of the 600,000 villages which have no electricity or the world's two billion people without access to drinking water have the right to ask science to find solutions adapted to their very meagre resources. Humanity also as the right to ask science to give priority to research into processes of global disruption and ways of coping with them. What's more, all citizens have the right to ask science to further our understanding of the mechanisms of inequality and exclusion which are gradually undermining peace and democracy.

One major purpose of this paper will be to see that the benefits of science go primarily to all those who have hitherto been unreached. Their conditions will only improve if they have access to the mighty power of science.

15

Democracy and the Market Economy

Today the idea of democracy is triumphant; the model is in principle embraced in most countries the world over. You may say that the very word democracy has been hailed and misused earlier in history. The most repressing and totalitarian regimes have tried to mask themselves as "real" or "peoples" democracies. What has happened, however, is a historical demasking of these false pretences.

What exactly do we mean by democracy? There is now a general agreement that democracy cannot be defined by purpose or policy or levels of mass mobilisation. It must be defined as a political system where different parties or individuals compete for power through regular free elections where all adult citizens have a vote. Moreover, a democracy must uphold certain basic human rights and well-defined freedoms which make the political process possible, and respect the opinion and integrity of the individual. No other definitions hold, and we should be careful when we talk about "real" democracy versus "formal" democracy. A society which in real life upholds the constitutional or formal democratic principles and which in practice applies the rights these principles imply, is by definition a democracy. A society with a beautiful-sounding constitution but where none or few of these rights are respected is certainly not a democracy.

Democratic Government No Guarantee for Equality

It is important to understand that democratic government does not necessarily mean good government in the sense that those in power make wise or well-considered decisions. Nor does it mean that conflicts inherent in the society are reduced to a minimum. Demands for democracy, social justice and a better life have historically gone hand in hand, but this does not mean that the establishment of a democratic system actually does lead to an improvement in social conditions or equality. It is also quite clear that some societies have a sort of outer shell of democracy but in reality, exclude large groups of people from having any political influence whatsoever. The actual differences in living conditions are so enormous and so entrenched that these people have no confidence at all in the political system even if it is democratic according to the definition. In these cases – for example in some Latin American countries – one can talk of a "masked hegemony with competing elites" where the outcome of struggles for power has little relevance for the masses. It is a sort of social and political half-authoritarian system – but disguised as a democracy – where the military often have a significant influence.

In the rhetoric of the day the terms market economy and democracy are used as if they were synonymous or atleast naturally emerging at the same time. But this is wrong—or atleast misleading. When the market economy or capitalism finally established itself in the 1800s and came to characterize modern industrial civilisation, democracy was at best in its infancy. In fact one could argue that democracy grew out of the contradictions and social dynamism inherent in the market economy of the capitalistic system. In this century we have a long list of terrifying and repressive regimes which have nevertheless upheld the virtues of a market economy. That some of these regimes have for ideological and security reason been hailed as bastions against communism, and also dignified members of the so-called free world does not transform them into democracies. In this company it is perhaps unnecessary to remind ourselves that the colonial system was assuredly not democratic, but was certainly based on capitalistic or market economic principles. It is the sad but irrefutable historical coupling between Western democracy, colonialism, and the plundering of resources in the name of the market economy which for understandable reasons meant that many of the leaders of national liberation movements looked for other models for the development of their young nations. In this connection it can be worth remembering what Nelson Mandela said soon after his release from 26 years of prison in the market economic but racist state of South Africa. "When we in ANC during 40 years struggled for democracy we were put in prison by the same people who are now telling us how we should behave to promote the democracy we have been rejected by all these years".

While we can see that a market economy does not automatically lead to democracy, a functioning democracy – as we have defined it – does seem to require some form of free economic system.

Democracy and Economic Freedom

Theoretically, it is conceivable that a political democracy could be combined with an economy totally controlled by the government – but experience has shown this to be very difficult. One could even argue that it is by definition impossible since democracy implies a certain freedom of economic choice and independent economic actors. A functioning democratic system presupposes what is now often referred to as a civil society – in practice, independent institutions, companies, organisations, the media etc., regulated by law but not subject to or controlled by those in power.

We must also see clearly that there are no unambiguous relations between economic growth, development and democracy. Democratic governments are neither very successful when it comes to structural reforms which may be to the disadvantage of important interests in the society, nor when it comes to welfare. The developing countries which have achieved the greatest success economically and socially over the last 20 years are the East Asian countries – which all have had various kinds of more are less authoritarian systems.

However, that does not mean that you can use these countries as models for the rest of the world. These is no globally valid link between an authoritarian form of regime and economic development, not even when development is defined only in terms of autocentric growth. Many social scientists have tried to find some systematic connection between what we call development or modernisation on the one hand, and the political system on the other – but all have failed.

It is also obvious that one of several prerequisites for economic growth and development is legitimate and reasonably well functioning government and governance. If the free market is to be a motor for development and improved welfare, and not just a meeting place for robber barons, the mafia and speculators, you must have a regulating and supportive state. If economic history teaches us anything, it is just this. Consider the astounding development in Germany after the war, or in Japan and the other East Asian countries some years later. There are many differences, but what they have in common is a well-functioning government apparatus with a long tradition.

Today we find ourselves in a historical situation where a large number of countries in the former communist states of Europe, in Africa, Asia and Latin America are at one and the same time trying to establish a new democratic system and new economic mechanisms. The situation is unique, and the intrinsic problems are unprecedented. Democracy as an idea has triumphed but in its practice it is in profound trouble. It is no exaggeration to talk of the crisis of democracy.

The former communist countries are certainly in crisis. As a by-product of the past regimes, there is an intensive suspicion of the political institutions, of the state and the parties- and in this way also the legitimacy of democracy and the ability of the politicians to deal with the fundamental problems of society has been undermined. The lack of a democratic tradition is not overcome from one day to the next.

Many of the developing countries have similar difficulties. The introduction of a multiparty system does not in itself mean that one can manage the conflicts and social problems in a democratic way.

Countries in Transition

Both in the East and the South countries are trying, at one and the same time, to change the political and economic system. When the whole society is convulsed by economic changes, and where peoples' living conditions fundamentally change, it is not easy to develop and maintain a political system based on compromise and respect, including respect for minorities.

As in previous history the deep crises of legitimacy and general frustration feed national and ethnical conflicts. These conflicts establish themselves in societies where the authoritarian system, economic crises and the breakdown of traditional values rob people of any kind of kinship other than ethnical.

We cannot avoid seeing disturbing signs of this crisis of democracy also in the so-called "established democracies" of the rich countries.

It is obvious that the state of democracy varies from country to country, as do the reasons for a feeling of dejection. But there are some similarities too.

The continuing and noticeable internationalisation limits the national freedom of political choice, available alternatives, and makes it more difficult for people to see the connection between "politics" and their actual living conditions. The governments are restrained by international economic events. The reaction of the stock exchange may be more important than that of the voters. The election results influence the stock exchange prices – but is it perhaps not also so that the stock exchanges, indirectly, also influence the elction results? People feel themselves to be the victims of major economic changes, but no one seems to be responsible and they themselves feel they have little chance of influencing the outcome. The absence of clearly identifiable alternatives between the larger political parties provides between the larger political parties provides opportunities for the populists and the extremists.

There is indeed reason to reflect on the lessons of the history of our turbulent and cruel century.

Priority for Growth

There is today much concern about the lack of resources for such urgent needs as the reconstruction of the East, a concerted attack on poverty and human development in the poorest countries, and environmental investments of all kinds. If the growth of world output returns to the levels of the 1980s, total output would grow by about one trillion dollars a year. There is, in fact, no other way to resolve the economic and political crises multiplying in the world community than to give priority to the restoration of growth.

We are certainly not at the end of history as someone has argued. We are rather at a dramatic turning point, a moment of many possibilities and many dangers. What we do now, for a few years ahead, may direct the future for several decades – like the dramatic and fateful years immediately after the second world war. All nations, all governments, have a responsibility. The rich world has a special responsibility, not just moral because of its enormous economic and political power.

16

A World in Transition

In a number of countries emerging from authoritarian rule, a new quest for the rule of law is being encouraged by popular pressure and globalization. If there is consensus on the need for a rule of law, the dynamics involved in success or failure in getting to such a goal are varied.

On the steppes of Central Asia in countries not so long ago under Stalinist rule, youths are being taught something unimaginable a decade ago: to defend their legal rights in the event of being stopped by police.

The passion for the principle, if not always the practice, of the rule of law is not coincidental. Much of the world has changed radically in recent years, moving from one-party authoritarian rule and command economics to multi-party politics or to free markets, and often both. Rule of law – a system in which, in theory and in practice, the law is binding on all persons as well as the government, a system which treats all equally. This was followed by the collapse of communist regimes in the Soviet bloc, and the shift towards more democratic political system and free market-oriented reform. Meanwhile, a number of states and territories in Africa and Asia, as a result of mass movements (Benin, Indonesia) or political evolution (Taiwan), are changing authoritarian or dictatorial regimes to more participatory and accountable systems.

Momentum for legal reform has come from growing popular pressure at national level, as well as from globalization, the catch-word not only for increasing trade and economic integration, but also for standards of civil conduct affecting ordinary people's basic rights. Membership in economic and political groupings and federations often brings international pressure for countries to protect basic rights through a rule of law.

While a set of laws in itself doesn't guarantee individuals rights—Nazi Germany and Stalinist Soviet Union had legal codes—a legal system equitably and tenaciously implemented is a cornerstone of democracy.

The Force of Tradition

If there is consensus on the need for a rule of law, the dynamics involved in success or failure in getting to such a goal are varied. First of all there is tradition. For instance, one legal scholar traces the absence of the rule of law in Russia back to the influence of Russian Orthodoxy on the legal consciousness and a tradition of absolutism uninterrupted for nearly half a millennium. On the other hand, a strong traditional sense of freedom can give impetus to the establishment of the rule of law after a period of dictatorship. Secondly, socio-economic factors play a part: in free markets, workers, business owners and investors all demand protection. Then there is the weberian analysis that as political leaderships move away from charismatic legitimacy based on great deeds—such as leading a revolution—they seek legitimacy elsewhere, including in effective government through the law.

In the 1960s and 1970s USAID (the U.S. Agency for International Development) and the Ford Foundation sponsored law reform programmes in the South directed more at economic and social development than political development. By and large they failed. Governments which passed the laws had little commitment to them. Sufficient resources were seldom provided for the implementation of the laws. Most projects overestimated the capacity of states to absorb new policies and institutions; for example there were not enough, or not enough well trained, lawyers, judges, administrators, who understood or were able to operate the new laws. Most projects underestimated the persistence of custom and customary practices.

17

Free Trade as Peacemaker

The Benefits of an Open World Trading System

Globalization by free trade according to the principles of the World Trade Organisation (WTO) offers the only realistic opportunity to integrate the world peacefully and in time to prevent a major disaster. The primacy of the economy over politics is the most important vehicle for a successful world domestic policy.

Since Adam Smith, traditional economic theory has on principle been well-disposed towards free trade. Free trade enables better use of the world's economic resources than does national protectionism. Countries can concentrate on their respective strength and draw from their trade partners the goods they need, but do not produce. But there have always been objections against free trade.

The international trade system has always been encumbered by disparate accusations of unfair competition. The fear that foreign competitors use unfair methods, such as dumping, as and is widespread. If one were to believe all the charges of dumping that are made, then international trade would have been completely destroyed long ago. Great restraint should be exercised with respect to allegations of dumping if one is interested in maintaining an interweaving of international economic activities.

The Free Trade Opposition Cloaks Itself in Dumping Charges

The modern form of the struggle against free trade cloaks

itself in the accusation of ecological dumping or social dumping. With this difficult subject matter, one should not make sweeping generalizations. These things also are not gone into the detail in what follows.

Environmental protection is an asset that every economy produces at the cost of other assets. The people's preferences for the asset of environmental protection probably varies from country to country. It is also completely legitimate and does not at all distort trade if the environmental provisions in line with the different national preferences – vary from country to country.

In the rich western European economic region, one should guard against a new form of cultural imperialism. It is not for this part of the world to impose its preferences for environmental assets on other countries, especially Third World countries. Free world trade brings not only economic advantages. Even more important is its contribution to lasting world peace.

In view of world population growth, every standstill in the movement towards a peaceful world society must be seen as a step backwards. We are compelled to run a race between the growing problems and the development of stable institutions to overcome them peacefully at global level. Economic history since the end of the Second World War shows clearly that free trade under the old GATT was of decisive importance for the prosperity of the industrialised nations.

The principle of help for self-help has nowhere been applied so consistently as on the free world market. In reverse, the examples of countries that cut themselves off from the world market show the disastrous consequences of the rigidity of a society which shuns the pressure of international competition.

Revolutionary Success of Open-Market Policies

The West's policy of open markets pursued since 1948 and reinforced since 1989 has led to a dynamism which, in the true meaning of the word, is revolutionary. More than half the world population now lives in countries with annual GDP growth rates of more than 5 per cent. Europe is not among that group, which may be why it also stands somewhat apart in its mentality.

Certainly, there also can be undesirable trends in free trade. There are no ideal systems; one must choose between imperfect potentialities. However, no realistically better substitute for the free trade system is in sight, not even with respect to the goals of a pacified world: an ecological sound world economy and a balance of global dimensions between the poor and the rich. An ideal government of philosopher kings armed with absolute power certainly could do some thing better than does free trade but such a government remains factitious. There are tangible and narrow limits to what the political system, whether democratic or not, can effect in a positive sense. This is how the structural conservatism of democratic and other political systems impedes the timely assertion of reforms necessary to achieve a world peace society.

The GATT was turned into the World Trade Organisation (WTO) a few years ago. Besides extending the free trade principle to services and additional agricultural sectors, the new agreement foresees above all the full inclusion of the Third World in the system. The agreement commits the industrialised nations to open their markets to developing and threshold countries.

Other important points are the strengthening and tightening of the dispute mediation process. Based on a system of relatively independent ad hoc panels, it permits complaints against WTO member countries for violations of the agreement. Thus, what is arising here is an effective global jurisdiction within the meaning of a peaceful world domestic policy.

Exclusion as Penalty

The decisive sanction mechanism of the WTO—which is not a specialist organization of the United Nations—is the threat of exclusion. Exclusion would deny the penalised country free access to the markets of WTO members on the basis of most-favoured nation status. This is a threat that requires no armed force, but is very affective. No country can still afford to do without the beneficial effects on prosperity that participation in international trade brings.

Thus, with the threat of denial of access to world markets for violating WTO rules, and the guarantee of a more or less fair competition for a country's own products for abiding by them, a non-military sanctions system has come into being. That is substantial progress on the path to a pacified world.

Certainly, this sanction system's sphere of influence is limited for the time being. Essentially, it will be used to assert the game rules of free trade. It offers no legal grounds for pressing other goals, such as on human rights. Attempting to expand it in this direction would for the foreseeable future put the entire system at risk.

In the current debate on globalization, the question arises of whether the world economic institutions should not be converted in this manner, that politics regains its autonomy, and that the primacy of politics can be restored. The critics of globalization point to the constraints to adjust which the world economy exercises on national or continental politics. However well this demand for the primacy of politics may be justified in philosophical terms, it virtually comes down to a demand for the ascendancy of the conservative principle.

Danger of a Slowed-Down World Integration

The demand for the primacy of politics is gaining strength from the desire to avoid the pressure to adjust which the dynamics of world events are exerting. It is today a conservative, and in fact a reactionary, longing for the (Utopian) return of the functioning European welfare state of two or three decades ago. If it were asserted, it would mean practically slowing down world integration. It would run dead against the goal of a world policy based on a desire for peace.

The present primacy of the economy over politics—in terms of the free movement of goods, services, and capital—is basically nothing more than the priority of the global principle over the provincial, the national principle. As such, it gives the principle of change pre-eminence over the principle of maintaining the status quo. What gives the primacy of the economy its legitimacy? Probably not the thought that world peace better be secured by this means. Its legitimation lies in the very indirect economic success that the free trade system delivers. For the reflective observer, the question remains of whether this legitimation is sufficient.

To answer this question, however, and particularly if one pleads for maintaining the ascendancy of the economy, it appears appropriate to outline the consequences that can be expected from further integration of the world economy. As can be seen today in East and Southeast Asia, the growth dynamics of the world economy will lead to a marked rise in the living standards of a large part of the Third World.

Do Not Exclude Poor Countries from the Competition

The global consequences of Asia's growth should not have been seen only negatively. While it also may mean, for example, a great burden on the global climate, it leads at the same time to an acceleration of the process of falling birth rates and thus to an earlier stabilization of the world population. Prosperity for the third World is so far the only realistic answer to the urgent problem of population growth. And free competition on the world market is the only reliable means of achieving this prosperity in the course of some decades.

Despite ecological sacrifice in the medium term, continuation of Third World growth is the only way to solve the long-term ecological problems. One also should not forget that only those who can eat their fill and have a roof over their heads are prepared to reflect on ecology and discuss it.

As for the rest, the balance between rich and poor is more acceptable when the poor become richer than when the rich become poorer. That applies also at the international level. The market and access to it are peaceful sanctions of the world economic system on the basis of free trade. Those who are hungry and have nothing more to lose are more of a danger to world peace than those who have eaten their fill. The ruse of covering up domestic problems by cross border military aggression will become less attractive to the degree that a country's own economy is integrated in the global economic system. The more countries are economically dependent on each other, the more unlikely it is that they will wage war on each other.

Globalization by free trade according to the principles of the WTO offers the only realistic opportunity to integrate the world peacefully and in time to prevent a major disaster. The primacy of the economy is the most important vehicle for a successful world domestic policy.

18

Aid Effectiveness as a Multi-Level Process

Parallel to the widerspread decrease of aid resources provided by donor countries to developing countries in recent years, debate and research on how to make aid more effective has become a major concern. Usually, it is suggested that decades of development assistance have at best produced marginal results in terms of improving development levels in the South. Little mention is made of donors' policy shortcomings and the negative impact of these on efforts aimed at reforming and redefining development cooperation in order to enhance aid effectiveness. The policy parameters and operating frameworks of existing aid policies continue to inhibit higher degrees of aid effectiveness. In many donor countries, opinion polls indicate waning public support for development aid.

Increasingly, the moral case for aid is called into question and deeper world market integration tends to be seen as the panacea to continued economic decline and social destabilisation in the South. Against this background, cooperation between donor and recipient actors is faced with a duel uphill struggle. First fewer resources can be mobilized to meet growing developmental needs. On the other hand, to organize and manage development policies and programmes in a result-oriented manner, grows more difficult. The threat of further aid cuts and of further drops of public support for providing aid become ever more real. A closer look at the organizational complexities and political constraints under which development cooperation is expected to perform effectively may help to improve current aid management approaches.

Towards Conceptual Clarity

At first sight, catchy definitions of what constitutes effective aid might appear attractive to use, in particular with regard to economic indicators, The term "aid effectiveness" is easily used in the same vein as "efficiency", "significance" or "impact" of aid. At times, obsession to measure and demonstrate the results of aid supported development processes can be observed among policy-makers and administrators on the donor side. Still the understanding of aid and its effectiveness as being part and parcel of a cooperation relationship between donor and recipient side parties, is scarcely embedded in practice. To determine how to make aid more effective requires more than a quick impact analysis of an individual and perhaps even isolated development project. Consequently, defining the concept of aid effectiveness needs to take into account at what levels cooperation is focused on. To strive for sustainable and effective modes of development cooperation will entail the need to combine recipient ownership of the development process with donor accountability concerns.

Performance expectations cannot be exclusively placed on the recipient while donor interests, their aid management systems and procedures remain unchanged.

An extended and more analytical process-oriented definition should take into account four main aspects of aid effectiveness:

(a) Effective aid must relate to the building and/or strengthening of in-country aid management capacity:

(b) To maximize the degree of aid effectiveness, local ownership of the aid process is essential: from setting of priorities through policy formulation and implementation on to the evaluation stages of the process;

(c) Increasing recipient side capabilities to take charge of aid relationship, will need to be combined with arrangements to meet legitimate donor accountability concerns;

(d) Aid effectiveness is a two-faceted objective: its realisation is equally dependent on increased

transparency of donor motives and on dropping of nondevelopmental, political and economic aid objectives of donors.

In addition a broader range of stakeholders in the aid relationship needs to be actively involved: extending beyond accountable government and implementing agencies, to include democratic institutions and organisations of civil society and of the private sector.

Applying any definition of aid effectiveness without disaggregating macro-economic data and taking into account country specificity will only lead to unhelpful generalisations about aid and its effectiveness. It would seem more appropriate to adopt working definitions against which to assess effectiveness of aid resources at a country-specific level. On such a basis one could expect to arrive at more reliable indicators of how well aid resources contribute to improving developmental standards and meeting exiting needs.

From Definition to Success – Key Requirements

Having reached agreement between the recipient and donor on what should constitute effectiveness of aid is only a starting point. Embarking on democratic, peaceful and participatory patterns of economic and social development must follow: to arrive at significant and lasting improvement in many of the least developed countries will be a long-term process. This being said, it is crucial to design and implement such forms of development cooperation which involve a wide range of recipient side actors, not only from the government side but also from civil society at large. Seen as a process of increasing inclusion of intended beneficiaries of aid, the commitment to decentralize as well as entrust aid and its management grows in importance.

To fully capture Third World development realities, policy frameworks inspired by neoliberalist-type of development concepts and theories are grossly inadequate. The views and positions on aid articulated in the World Bank and the IMF, or in many if not most bilateral aid administrations in OECD countries, represent only one side of today's international cooperation, namely, the donor side. The major weakness to point out with respect to this locus of debate, is a profound underrepresentation if not even a total absence of recipient experiences and perceptions on aid in general and on its effectiveness in particular. There should be little doubt that ignoring to not actively identifying and involving such perceptions, leads to strongly donor-driven aid.

To circumvent recipient side insights and views on strengths and weaknesses of aid strategies and mechanisms, will result in limited local commitment and sense of ownership over the aid process. Mutual decision-making between donors and recipients remains a rare policy approach. Aid procedures that are based on local management and less control-oriented donor roles in the aid process are still exceptions in development cooperation.

Structurally, in terms of the policy environment within which development aid is expected to function, the overriding policy framework is generally based on structural adjustment policies (SAP). But the underlying conclusion made by proponents of SAPs that these policies induce aid effectiveness, has yet to be proven valid. It must suffice at this point to emphasize that there is no a priori relationship between world market integration under structural adjustment and sustainable development in poor countries. Aid to these countries which is solely intended to reinforce fundamentally uneven and unequal patterns of world market integration should at be scrutinized critically.

Some central issues need to be addressed in the course of improving aid and its effectiveness:

- institutional dimensions of aid relationships require strong policy-attention, both on the donor and the recipient side;
- capacities to effectively identify and formulate aid priorities need to be strengthened in recipient countries;
- local capacities to sustain reform efforts must be reinforced.

Levels of Intervention

If the design of aid and the terms upon which it is provided to a developing country are largely determined by the donor, the aid relationship can be characterised as essentially hierarchical. Recipient side views will rarely surface, as they are either not identified, or not well formulated. Possibilities of recipient-led development strategies can be limited. Unless scope is provided to the recipient side actors to assume responsibilities, aid effectiveness is likely to remain low or fluctuating, and the sustainability of donor aid efforts will remain doubtful.

National planning processes and courses of national development in recipient countries should be seen as most effective where they are led under local responsibility and control. To arrive at this ideal situation, gaps need to be reduced and closed at the various intervention levels.

Donor aid resources provide valuable support for this process. Their effectiveness in meeting long-term objective of aid will need to be assessed on the basis of how well they perform at the different levels. Individual donors will expectedly perform differently at the various levels. What will prove to be the ultimate test for effectiveness is how well the donor aid performance accomplishes the broader objectives of development cooperation and how well it includes sustainable results.

In the analytical frameworks outlined here, development cooperation would seem to be confronted with the effectiveness gaps at the.

- structural level: International trade and investment patterns, debt problems and world market integration process appear as long-term constraining factors upon aid and its effectiveness;
- at the policy level, dialogue and partnership in development cooperation are instrumental factors in recluding planning and coordination gaps with regard to policy analysis and formulation;
- the institutional level is where pertinent capacity gaps exist: capacity development efforts of donors and technical assistance measures play an important role in addressing weaknesses in aid effectiveness within a country's institutional setting;

- finally, at the level of aid projects (programmes), it is generally the lack of sustainability of aid interventions which causes development activities to falter once donor support decreases or stops. In addition to technical cooperation, financial and material inputs serve to maintain project momentum and goal realisation: the issue of how to develop local capacity sufficiently in order for indigenous organisations to continue project activities initially supported by donor aid, remains the most important issue to address at this level.

Fostering Aid Effectiveness

Donor and recipient development efforts are too often isolated from one another, or poorly coordinated. They fail to address managerial and implementation bottlenecks. Cross-sectorial linkages, as well as interdisciplinary approaches to aid problems are only slowly gaining ground. It is increasingly obvious that decisions on aid issues are subjected to concerns outside of the responsible ministry: finance ministers, and unfortunately even defense ministers have a strong say in how much aid is to be provided, where it is to be concentrated and under what terms to be utilised. Inside of recipient countries, large portions of national budgets are allocated to non-development priorities with little or no impact on alleviating urgent poverty problems.

Development cooperation may make the biggest impact and be executed most effectively where donors and recipients agree upon multi-level aid strategies. To give an example: building a road to a remote rural area may well be done in an effective project manner: it is equally important to have a functioning transport authority in place to ensure maintenance of the roads. If this authority operates within a nationally defined infrastructure policy, best in accord with national trade and investment priorities, then the effectiveness of the project-level road building programme has a good chance of being high.

Institutional changes to set the stage for a profound reform process in development cooperation are needed. Re-prioritising national budgets to reflect identified in country development needs may be one step. Setting up policy evaluation and formulation units can be complimentary measures. Deregulating markets and investment rules may serve to please donors, but dumping of cheap products which strangle local production efforts may easily result. Regional cooperation, including intensified South-South cooperation can provide some counterbalance. There are only a few areas where changes in the current system of development cooperation can occur, with a view to better manage the complexities of aid and the social, cultural, economic and political backgrounds against which they take place. The will and commitment to take policy action in both donor and recipient countries, through the broadest range of stakeholders and institutions as possible, will be the test for genuine efforts at improving development relations between North and South and organising cooperation effectively.

19

Health Care Relief in Conflict Situations

What can We Learn from the Food Relief Experience?

Conflicts and war occur in many of the poorest nations where populations already suffer from severe ill-health. War leads to an increase in disease and to a worsening of the already fragile condition of populations. Health care itself becomes a victim of conflict. Many deaths which occur during these emergencies are not discretely related to the conflict itself but are the result of lacking access to public health services. Furthermore, conflict contributes to the deterioration of already pre-existing structural weaknesses of the health care system. An example is the period of internal conflict in Uganda (1970-1986) when health services declined in the aftermath of the war due to the impact of foreign assistance and the planning vacuum in which the activities took place.

The Impact of Conflict on Health Care

Conflict and civil strife may lead to a major disruption of health services. This is not only a result of physical destruction but also of finding shortages since national governments increase spending on military activities. Casualties increase the demand for curative services which can divert already limited resources from preventive care.

In the case of the Sudanese civil war a large majority of health professionals was forced to abandon rural health services and left for urban areas or neighbouring countries in order to find new employment. Entire preventive health services such as immunization as well as water and sanitation projects collapsed leaving the population exposed to infectious diseases and epidemics. In urban areas, the gap in pubic health care provision is sometimes filled with the expansion of private services. In rural areas, private sector involvement in health care is rather marginal, apart from some omission hospitals or pharmacies. Therefore the non-formal health care sector often makes a substantial contribution towards health care.

With the rise of internal conflicts in Africa, more people suffer from emergency situations. This also increases the influence and impact of international donors. External assistance nowadays accounts for more than 25 per cent of government health expenditure in sub-Saharan Africa.

The size of donor involvement reflects the power of international agencies to control the policy domains. Countries in conflict or post-conflict situations are under pressure to 'rescue' their health systems and accept global policies in exchange for aid assistance and relief.

However, in the period after 1991, donor organisations tended to increase their expenditures for high profile humanitarian operations rather than ordinary development activities. This shift may reflect the increasing influence of media covering some of the conflicts. Too often, organizations intervene with ad-hoc assistance without sufficient consultation at local level.

Donors' Perceptions in Designing Relief Interventions

Today, in many parts of sub-Saharan Africa development assistance has virtually collapsed and has been substituted by relief assistance. The problem is that relief interventions are based on a Western construction of reality, reflecting what is desirable and necessary in times of conflict. Most interventions therefore stress physical and material needs, presuming that the social aspect of food and health is not an immediate issue to address.

The question which arises here is on whose views and perceptions these needs are based? While donor interests may be guided from the perspective of ill-health, the recipient government may be concerned with the collapse of the economy. However, any intervention needs to take into account that local knowledge and practices are shaped by state interest as well as power relationships. The common belief that health care systems always collapse due to conflict is sometimes mistaken. Considering the fact that toady's internal conflicts are often fragmented, conflicts do not necessarily result in a breakdown of the health care delivery system.

Donors tend to respond with a 'package' approach and developing countries' ministries of health increasingly play a symbolic role. The evidence suggests that international organizations tend to create verticle programmes which undermine national public health programmes. Foreign interventions are technically sophisticated and reorienting health are towards a more curative approach. Too little attention is given to strengthen the health care system within its own limits, providing more appropriate technology, drugs and emphasizing the training of local health staff.

Another vital issue concerns the existence of already fragile health information systems. Agencies tend to bring their own systems which lead to further fragmentation. The local perspective on what are the 'basic needs' in physical and social health are usually not considered. Health relief interventions do not recognize the potential of the communities and the non-formal health sector such as healers and traditional midwifes in supporting and maintaining health care sector presents a substantial contribution towards health. It is not the question between choosing either allopathic or traditional services, it is more the decision which kind of illness will be best treated by which practitioner. There is a need in further exploring the role of this sector particularly since this is sometimes the only service available for certain populations.

Responding to Local Needs

More community-based public health interventions could be vital to reduce mortality and morbidity. For example in Somalia during the 1992 war and famine high mortality rates due to measles and diarrhoea could have been prevented by involving the communities in primary health care activities such as immunization and nutrition improvement.

In the African context Tigray is an example where health services had been sustained and partially expanded during the civil war against the Ethiopian government. Local government structures called baitos promoting social and economic development. Baitos encouraged communities to establish revolving funds for drugs and medical equipment. It actually functioned as an early type of community financing system.

As mentioned above, the challenge in changing health care relief strategies is to overcome the approach of short-term interventions, particularly in a changing conflict environment where conflicts are complex and interruptions are no longer short-term.

Therefore interventions need to be linked with the process of conflict resolution to avoid health care or food aid being used by politically dominant groups.

Food Relief in Conflict Situations

Food interventions have both a survival and a production function. For example, food-for-work may be part of an income programme or food aid can be monetized to generate local currency. However, food aid has to be seen beyond the objective to fulfil nutritional goals, it also defines relationships between social groups in regard to food accessibility and how food is shared. Food aid is aiming to meet people's basic food requirements and minimising risk and severity of disease by complementing services such as basic health care.

In more stable political conditions where free food aid is given it presents an income transfer by releasing income which normally is spent on food. However, in conflict situations food relief frequently becomes part of the dynamics of conflict such in the case of Sudan where it is used to sustain the struggle between the North and the South without resolving it. Furthermore, the military attack food supplies in the fight against rebels who depend on the support from the communities.

Health is also a matter of food security. When food insecurity coincides with conflict situations, health and survival are threatened. Food security provides some concepts on how and why vulnerable households manage to survive in periods of hardship (coping strategies).

Coping Strategies in African Trouble Zones

Today, most conflicts in Africa such as the one in the Great Lake Region, Angola or Congo cause major problems of food insecurity. They are linked to the civil wars which produce substantial social disruption as a result of massive population movements. The analysis of coping strategies showed that household respond to these conflict situations by eating less, selling livestock and land, or trying to find new sources of income.

In some emergency situations, however, such coping mechanisms may fail. In the case of the war in Mozambique food aid was vital since coping strategies were limited and people had to sell all their assets which was particularly true for internationally displaced persons and refugees.

It has been argued that food relief bypasses local structures in favour of those qualifying on a nutrition status criterion, decided by international organizations, or it may attract populations to refugee camps to receive free food rations and thereby undermines local production. In the case of Rwanda food aid was targeted at the internally displaced and left out the local population. This can be due to donor bias in needs assessment.

Food scarcity is not always so result of civil war but its creation may be rather a political objective. An example is food relief manipulated by local elites and the military like in the case of Sudan. It can be summarized that generally relief operations often bear the risk of fueling the process of instability and violence rather than helping to contain the situation.

Lessons from Food Relief for the Health Sector?

Through the experience of food relief in recent civil wars such as Sudan, Somalia, Mozambique etc., there has been an increasing awareness of the economic and political context in which operations take place. Like food relief, health care is a political tool which can, if not properly targeted, undermine people access to health care services. While food production is linked to food security, it is more difficult to identify factors leading to self-sufficiency in health care.

As mentioned above, food aid is aiming to insure survival. It also has an economic aspect, protecting household assets. Health care relief is targeted to assure immediate physical survival based on the importance of social health. Unfortunately, curative interventions hardly consider the socio-cultural dimension of health. Therefore it would be beneficial if health care interventions consider local norms and traditions. Interventions should be compatible and complement local health programmes. The emphasis should be on strengthening formal and non-formal health institutions both in service provision and training.

In food relief, distribution and needs assessment identification are controversial issues for discussion. While the programme design is shaped by donor perceptions, the actual programmes are influenced by the priorities of some powerful leaders as well as the socio-economic and political context.

Health care interventions need to analyse these issues in the context of economic and political systems in order to identify the most vulnerable groups, for example populations living in areas which are more, operations require a stronger involvement of communities both as users and active participants to carry out and maintain public health programmes.

There is a need for a new concept to be designed which applies to chronic emergencies. In the absence of a policy framework, guidelines need to be developed in order to overcome the inconsistency in planning and implementation. Donors need to change their assumptions on which they plan their health relief responses. A starting point in improving the efficiency of these operations is to provide institutional support to local authorities and organisations and involve them in planning and implementation.

20

Law and Social Justice

Law reform in the service of democracy must find ways of protecting the vulnerable. Legal reform and "good governance" have vaulted to the top of the development agenda. International financial institutions and influential donors continually stress the importance of the rule of law, a healthy regulatory environment and strong and consistent enforcement of rights to successful economic development. In the new world order, the state's role is to facilitate private activity rather than guarantee the welfare of its citizens.

But there is growing concern that market reforms and globalization are connected to great social stratification and economic inequality. What is often overlooked is that legal reform may enhance rather than alleviate this stratification and inequality.

It is important to see legal reform as a key part of a broader set of policy, legislative and institutional reforms which are designed to create not simply rule – and norm-based societies but particular types of market economies. There are no "free" markets; functioning markets depend upon a legal infrastructure and a commitment to the rule of law. The growing interest in legal reform indicates nothing if not the widespread recognition of this fact.

Tradeoffs between Efficiency and Equity

Respecting the rule of law and protecting rights however does not mean that there is any one best set of laws, even in market economy. Yet legal reform projects in developing and transitional countries have become inseparably associated with the idea of single, optimal path or model. Current projects emphasize strong protection for property rights, the consistent enforcement of contracts and, increasingly, financial sector regulation as the foundation of an investor-friendly legal infrastructure. At the same time, states in transition to markets have been discouraged from adopting or retaining "excessive" regulations, including protective labour market policies that might impede growth and efficiency.

Market-oriented legal reforms can affect the fortunes of different groups in at least three different ways. The first is through the types of reforms that are implemented. Because legal reforms allocate rights and entitlements, different rule structures may well benefit different groups in different ways. In some instances, there may be tradeoffs between efficiency and equity. Strong property rights will protect owners and entrepreneurs but may contribute to the disadvantage of renters and works; environmental and consumer protection laws protect the public at large but impose costs on businesses.

Second, people can be affected by the absence of particular laws. Labour standards and laws authorizing collective bargaining, for example, have been crucial in industrialized societies. If they are weak or missing as they are in many developing countries, or if they are indefinitely postponed because priority is given to implementing laws and regulations which facilitate economic transactions, vast numbers of people can find themselves worse off than they need be in the market for labour. Particular groups may also be harmed. Women with caregiving obligations are likely to be systematically disadvantaged and shut out of better work opportunities without market regulations which ensure that part of these costs are borne by others. This is especially likely where social programmes and subsidies are reduced or eliminated at the same time, as has occurred in many parts of the world.

Open Debate

Finally, where legal reforms follow a "standard form" or are designed by experts from afar, a common experience in transitional states, the risk is that local history and priorities will be ignored or displaced and democratic control over decisions about basic social organization is weakened. To avoid aggravating inequality and worsening the position of those who are frequently already vulnerable in the reform process, three conditions need to be met.

First, conflicts of interests—between workers and entrepreneurs, for example—as well as the necessary tradeoffs that legal reforms often entail should be acknowledged openly, rather than hidden behind the veil of efficiency. This will allow countries to debate more openly the political and distributive choices that legal reforms involve. Second, donor countries and international financial institutions need to rethink the position that state "intervention" is usually or necessarily the enemy of economic development. Third, developing states need much more space, indeed they should be actively encouraged, to accommodate distributive, equity and social concerns not only through social programmes and transfers but through the processes of legal and regulatory reform as well. This would allow greater attention to labour market concerns and environmental protection as well as to poverty alleviation and gender, racial and ethnic equity.

21

A Crucial Encounter

Genetic tests and treatments must not be allowed to create new forms of discrimination between those who, for whatever reason, can or want to take advantage of them, and those who cannot, mostly for lack of money. If a scientific discovery can form the basis of a technology, then it is highly probable that the technology will eventually be applied. Today this lesson of history is causing anxiety among politicians, scientists and public opinion concerned about the current far-reaching developments in biotechnologies.

It is now possible to penetrate to the very essence of living things as a result of spectacular scientific advances that are gradually revealing the innermost mechanisms of life. The technologies based on this field of knowledge offer humanity for the first time astonishing powers to revolutionize the process of creating and developing human beings and, ultimately, the human species. Technically speaking, these breakthroughs could lead to the revival, in even more effective guises, of eugenic practices we hoped had been buried forever. Fortunately, this nightmare scenario seems highly unlikely.

But history also shows that new technologies are rarely applied without a framework of rules and procedures designed to ensure that they are beneficially used. Human progress has always been driven by the winds of freedom, including freedom of enquiry and initiative, but human beings have always tried to head in the right direction and to respect certain limits. The biologists have done their work: they have sown the seeds of vast possibilities. Now it is up to society to make sure that only the benefits are harvested. The biotechnology revolution beckons humanity to a crucial encounter between science and ethics.

Where human reproduction is concerned, as with technology in general, we must be guided by respect for three basic and interdependent principles dignity, freedom and solidarity.

For human dignity to be respected, each person must be regarded as unique. This position has far-reaching consequences for human procreation. First of all, it rules out cloning as a means of reproduction because this technique, which is almost upon us, involves genetically "duplicating" an existing person. More generally, predetermining the basic characteristics of a future person, notably trying to enhance their future physical or mental capacities, violates the very essence of human individuality. This kind of engineering would end up by depriving individuals of that which is theirs alone – the mysterious processes whereby their unique genetic heritage emerges and interacts in its own unique way with their environment.

Advances in prenatal scanning and testing techniques may confront parents with grave new decisions. The danger is that various kinds of pressures or even regulations will develop which only allow "genetically correct" people to be born. This would be totally unacceptable. No authority – be it political, social or economic – should be able to enact such a "genetic order", still less impose it.

So increasing emphasis must be laid on solidarity. Genetic tests and treatments must not be allowed to create new forms of discrimination between those who, for whatever reason, can or want to take advantage of them, and those who cannot mostly for lack of money.

The risk of uncontrolled, unmonitored genetic engineering increasingly looms over us. But we are starting to see the emergence of a new "responsible" form of genetic engineering in which the power of science is subjected to the power of ethics an ethics that benefits everyone, jot just a few, and looks towards future generations, not just short-term interests.

22

Measuring Population's Impact

There is no easy way to measure the overall impact of human activities on the environment. Nevertheless, several approaches have been developed as follows:

Environmental Resource Accounting

Environmental resource accounting attempts to place an economic value on "environmental goods and services" used—natural resources that conventionally have been regarded as free and used in common. These include unpolluted freshwater clear air, ocean life, forests, and wetlands.

Some economists argue that the value of environmental goods and services should be incorporated into estimates of gross manufactured capital, which depreciates in value overtime, environmental capital (such as forests, fisheries, and unpolluted air and water) currently is not considered to depreciate, and no charge is made against current income as it is used. A country could exhaust its mineral resources, cut down its forests, erode its soils, pollute its acquires, and hunt its wildlife and fisheries to extinction, but measured income would not be affected as these natural assets disappeared.

If natural resources were valued in the same way that manufactured assets are valued, it might help economies learn to use them more efficiently and to conserve them in order to assure continued use in the future. Such valuations also might help indicate the economic benefits of protecting the environment, as well as the ecological benefits. In other terms, instead of continuing to draw down their "environmental capital" until it is gone, economies could begin to live on its interest, maintaining the capital for use indefinitely in the future.

I=P x A x T

The equation I=P x A x T represents another effort to describe the overall impact of humanity on the environment. In the equation:

- I is environmental impact
- P is population (including size, growth, and distribution)
- A is the level of affluence (consumption per capita), and
- T is the level of technology.

Despite its limitations—for instance, inability to assign actual values to each component or to depict changes in the factors over time—the equation is valuable. In particular, it emphasizes that developing countries with large and rapidly growing populations affect the environment, even though their levels of affluence may be low, while at the same time countries in the developed world with little or no population growth have a substantial environmental impact because consumption per capital is so high.

The equation makes clear that slowing population growth is a key part of any strategy to reduce humanity's impact on the environment. For example, even if per capita resource consumption (A) declined or technologies (T) improved enough to reduce the environmental impact (I) of humanity by 10%, this gain would be wiped out in less than a decade because world population (P) is growing at over 1% per year since per capita consumption of resources is expected to increase as living standards rise, protecting the environment requires more efficient production technologies, less waste, and ultimately a stable world population size.

Ecological Footprints of Nations

Everybody has an impact on the Earth, because they consume the products and services of nature. Their ecological impact corresponds to the amount of nature they occupy to keep them going. In other words, we calculate the 'ecological footprints' of these countries.

Carrying Capacity

The term "carrying capacity" refers to the number of people the earth can support. Logically, population growth must stop at some point, or the earth would become overcrowded and its resources eventually would be depleted. But what is this maximum human population?

This question has been debated since 1798, when English economist Thomas Malhus predicted that population growth inevitably would outstrip the food and water supply at some point. Since the estimates of carrying capacity have varied a great deal depending on what assumptions are made out technology, consumption levels, and other factors that are not easily forecast. Some have even argued that the earth's carrying capacity may already have been exceeded in sense that the world consumed at the rate that Americans and Western Europeans consume.

While nobody can know how many people the earth could support, few would want to find out the hard way —by reaching this theoretical limit. Calculate the maximum number of people who could exist on earth seems less important than determining how resources can be used wisely and managed sustainably to improve living standard without eventually destroying the natural environment that supports life itself.

23

Social Summit

The issues which the World Summit for Social Development is called upon to tackle—poverty, unemployment and social exclusion – are among the most critical of our time. In spite of the considerable material gains that have been registered during the past 50 years, in spite of the tremendous technological progress that has been achieved, in spite of the more favourable international political climate resulting from the end of the Cold War, poverty continues to grow and inequalities continue to widen.

Over a billion people are living in poverty. Some 30 per cent of the world's labour force is not productively employed. Many societies are being torn apart by racial, ethnic and religious conflict and intolerance. These social issues constitute the greatest threat to peace, stability and prosperity in today's world and need to be tackled with urgency and determination by policy-makers at the highest level.

If the Social Summit is to leave its mark on history, it will have to be more than a ritual gathering of Heads of State and Government. It will have to result in a perceptible improvement in the social situation in countries throughout the world – in the short and medium term, not in some indefinite future.

It will have to acknowledge that these are truly global problems which require global action in a world characterized by growing inter-dependence. National action, important though it is, will no longer suffice.

A collective commitment by all the nations of the world to the goal of full, productive and freely chosen employment is of crucial importance to the work of the Summit. It is only through their labour that individuals can contribute to the creation of wealth for their benefit and for society as a whole, can lift themselves out of poverty, can be integrated into social, economic, political and cultural life.

In view of the growing globalization of the economy – characterized by freer trade and capital movements and greater exposure to world market forces – no country acting alone can achieve full employment without a favourable international economic environment. But we also expect emphasis to be laid on appropriate national strategies: macro-economic policies, industrial and agricultural policies which favour the growth of job-creating enterprises, education and training policies, active labour market policies which enable individuals and enterprises to adjust rapidly to changing market conditions, and policies which grant all citizens equal access to employment, education and training.

We expect the Summit to commit all nations to take special action to improve the employment prospects of those who are in a particularly disadvantaged position in the labour market—young people, disabled people, ethnic minorities. We expect it to recognize the special contribution made by women to the general welfare – through their role in the family and in production – and to end the pervasive discrimination and inequality still suffered by women.

We expect the Summit to emphasize that full employment does not mean the creation of any sort of jobs, but of high-quality jobs. We expect it to recognize the basic rights of workers – freedom of association and collective bargaining, freedom from forced labour, equal pay for work of equal value, equality of opportunity in employment and occupation.

We expect it to recognize the importance of measures which provide social security and minimum wages, which protect against arbitrary dismissal and which ensure occupational safety and health.

However balance has to be struck between the need for adequate protection and incomes on the one hand and need not to price workers and enterprises out of the market on the other. Realistic and socially acceptable solutions can only be found if those most directly concerned – employers and workers – are fully involved in decision-making on these matters through organizations of their own choosing.

Finally, and perhaps most important of all, we expect the Summit to determine the institutional framework and mechanisms for following up its conclusions. There is no point in adopting fine-sounding commitments and policies if there is no means of ensuring and monitoring follow-up action.

If the Summit fails to establish such a framework, its conclusions could well prove to be a hollow text that will quickly be forgotten.

24

Food Security: Availability and Access to Food

The world food situation has never been better. Enough food is being produced today that, if it were evenly distributed, no one should have to go hungry. World food production is increasing faster than population growth: per capita production increased by 5 per cent during the 1980s. Real food prices are at historic lows and have been declining for some time now. Yields of major cereals have more than doubled in the past three decades. These trends have contributed to complacency in some quarters regarding the world food situation.

Yet, more than 700 million people in the developing world do not have access to sufficient food to lead healthy and productive lives. More than 180 million children are underweight. Diseases of hunger and malnutrition are widespread. The desire to satisfy food needs has, in combination with increasing population densities and inadequate agricultural intensification, led to much degradation of environmentally fragile lands, such as forests and steep hillsides.

Over the next 20-30 years, farmers and policy-makers in developing countries will be challenged to provide food at affordable prices for almost 100 million more people every year – the largest annual population increase in history. Moreover, they will have to increase food production from more productive use of the land and without further degradation of natural resources: area expansion is no longer a feasible option in most of the world.

What future food security will look like depends not on exogenous factors over which we have no control but on the decisions and actions taken by the major players: households, private - and public-sector agencies, governments, and the international community. If we continue to act as we have in the 1980s and early 1990s, more people will suffer from food insecurity, it will be because some or all of these players failed to act in an appropriate and timely manner.

Feeding the World: Availability and Access to Food

There is enough food in the world today to feed everyone, if it were evenly distributed. Availability of daily food energy per capita in the developing countries as a whole increased by 0.7 per cent per year during the 1980s.

Twenty-five developing countries, incluidng about half of the African countries, were unable to assure sufficient food energy (2,200 calories per person per day) for their populations at the end of the 1980s even if available food energy were evenly distributed within each country. This is down from 45 countries at the end of the 1970s.

However, available food is neither evenly distributed nor fully consumed. Availability of enough food at global, regional, or national levels does not necessarily mean that everyone is well fed. For people to be food secure - that is, to have access at all times to the food required for healthy and productive life—there must be both availability of food and access to food. Access to food by households (and individuals) is conditioned by poverty: the poor usually lack adequate means to secure access to food.

Over 1.1 billion people in developing countries were living in poverty in 1993, more than 500 million in conditions of extreme poverty. South Asia is the home of about 50 per cent of the developing world's poor—more than 500 million people. Another 15 per cent are found an East Asia, 19 per cent is Sub-Saharan Africa, and 10 per cent in Latin America and the Caribbean. The prevalence of poverty (the proportion of each region's population that is poor) is very high—about 50 per cent - in South Asia as well as in Sub-Saharan Africa.

Today there are more then 700 million people who do not have access to sufficient food to meet their needs for a healthy and productive life; they often go hungry. Adults and children also suffer from diseases associated with hunger and poverty. For almost one-fifth of the total population of developing countries to be chronically hungry tarnishes the image of a world that is now considered food-secure because it produces enough food.

Great progress has been made in meeting food needs during the last 30 years. For instance, the number of underfed people declined from an estimated 976 million in 1974-76 to 786 million in late 1980s. But the problem is far from solved. Keeping up with increasing needs and demands due to population growth, income increases, and dietary changes are themselves a formidable challenge.

Hunger and food insecurity have a significant effect on health and nutrition of both adults and children. They can lead to growth failure in children. About 184 million preschool children in developing countries were underweight in 1994. About 55 per cent of these underweight children were found in South Asia and another 16 per cent of Sub-Saharan Africa. The proportion of children that are underweight is higher in South Asia (almost 60 per cent), but it is also significant in Sub-Saharan Africa (30 per cent) and Southeast Asia (31 per cent). It is worrisome that the number of underweight children in Sub-Saharan Africa during the 1980s from 20 million to 28 million is particularly striking.

In addition to energy deficiencies, micro nutrient deficiencies are also widespread in the developing world. About 14 million preschool children (under the age of five years) have eye damage as a result of vitamin A deficiency. Ten million of these children are found in Southeast Asia. Between 250,000 and 500,000 preschool children go blind each year due to vitamin A deficiency, two-thirds of these children die within months of going blind. Many more children are mildly affected. Recent research has shown that even mild deficiencies can increase mortality significantly. Vitamin A deficiencies are closely linked to diet, which can be influenced by agricultural research and policy.

Iron deficiency affects about 1 billion people in the world, particularly children and women of reproductive age. Iron deficiency leads to anemia, which if not checked can diminish learning capacity and increase morbidity and morality. In the developing countries, about 370 million women between 15 and 49 years of age—42 per cent of this population group—were anemic in the 1980s. Almost one-half were in South Asia. And there are tentative indications from South Asia and sub-Saharan Africa that the prevalence of anemia is rising in non-pregnant adult women of reproductive ages.

In sub-Saharan Africa, this trend is undoubtedly associated with deterioration in general standards of living, including increased poverty and food insecurity. Anemia partly arises from diets insufficient in iron, which again could be addressed through agricultural research and policy. For example, a possible reason why iron deficiency and anemia are going up in South Asia may lie in the decrease in production of iron-rich pulses during that same period, which in part reflects the larger research input into competing crops such as wheat in South Asia. This emphasizes the importance of considering the effects of diet and thus on health and nutrition in setting research priorities for yield-increasing research.

South Asia is the home of about half of the developing world's hungry and food-insecure people, but this population group is growing rapidly in sub-Saharan Africa. Much of the poverty and food insecurity is in rural areas, mainly in low-potential areas such as arid zones, but urban poverty is also growing rapidly.

Four Key Factors will Influence Future Food Production and Consumption

Global and regional food production and consumption during the next 10-20 years will be influenced by a large number of factors. Changes in the following four sets of factors are likely to be particularly important:

1. Economic growth and economic policies.
2. Population growth and urbanization
3. Rural infrastructure, agricultural production technology, and access to modern inputs, and

4. Natural resource management and environmental considerations.

The expected impact of each of these factors on future food production and consumption is considerable.

Economic Growth and Economic Policies

Economic growth must resume in the developing world, especially in sub-Saharan Africa. To support such growth, it is critical to

- complete structural adjustment and economic reforms;
- remove external barriers to growth such as trade distortions and subsidies in developed countries;
- liberalize trade and remove market distortions;
- enhance access by the poor to land, capital, and technology;
- expand investment in rural infrastructure, health, education, and agricultural research and technology;
- facilitate and sustain ability in agricultural production, and
- reverse the decline in international assistance to agriculture.

Growth in real per capita income during the 1980s was disappointing for developing countries as a whole. However, the low average rate of growth covers large variations among regions. The high rates of economic growth in Asia are expected to continue through the 1990s, while incomes in sub-Saharan Africa are expected to keep pace with population growth.

Future economic growth depends on internal policies as well as on the international policies as well as on the international environment. The extent to which current structural adjustment and economic reforms in Latin America, sub-Saharan Africa, the Commonwealth of Independent States (CIS), Eastern Europe, and selected countries in Asia and the Middle East are carried to successful completion at an appropriate speed and sequence is of paramount importance for future economic growth in those countries.

Closely related to this issue is the question of the most appropriate role of the state in a market–oriented economy with inappropriate institutions, poor infrastructure, and insufficient experience by the private sector in dealing effectively in a competitive market environment. Overreaction to past failures such as excessive and inappropriate state intervention may cause governments to take on a passive role where intervention is needed to assure that the markets function effectively and to deal with outside influences on the economy.

Future economic growth will also depend on the international trade environment, including trade distortions by developed countries, and access to external aid. Import restrictions for agricultural and nonagricultural products in Japan, the European Union, and the United States, along with domestic agricultural subsidies and implicit and explicit export subsidies for agricultural products, are of particular concern.

Population Growth and Urbanization

If progress in economic growth is not to be undermined by rapid population growth and excessive urbanization, effective population and migration policies are necessary to complement growth-oriented policies. Such policies must focus on

- universal access to family planning information and technology; and
- incentives to reduce rural-urban migration, such as provision of employment in rural areas and stimulation of agricultural and non-agricultural growth in rural areas.

Although the annual growth rate is falling for the world as a whole, the population increase during the next 20-30 years, of slightly less than 100 million people a year, will be the largest ever. Approximately 97 per cent of this increase is projected to occur in the Third World, with Africa alone accounting for 34 per cent of the growth. Thus although reductions is annual population growth rates have begun to occur in Asia and Latin America, they are insufficient to counter the absolute increases. Population growth rates of these magnitudes will greatly increase the need for food and other basic necessities.

Rural Infrastructure, Agricultural Production Technology, and Access to Modern Inputs

Continued progress in all three of these areas is critical to future food security

- Resources must be committed to infrastructure construction and maintenance. Labour-intensive public works programmes are a viable mechanism for building roads, reforesting areas, and engaging in soil conservation projects, while creating employment and income in rural areas.
- International and national agricultural research must continue to develop yield-enhancing production technology, especially in maize, millet, and other crops, as well as build tolerance or resistance in crops to pests and adverse climatic conditions.
- Farmer access to modern inputs must be facilitated through provision of credit and technical assistance. Inputs must be made available to all farmers on time and in required amounts.

The importance of investments in rural infrastructure within the context of rapid urbanization has already been established. Even without rapid urban growth, however, such investments are needed in many developing countries, particularly the poorest ones, to facilitate agricultural and rural development. Improved rural infrastructure enhances access to export markets, modern production inputs, and consumer goods. It reduces marketing costs, promotes exchange between intracountry markets, reduces spatial and temporal price distortions, and, in general, increases efficiency in production and marketing.

However, while essential, effective rural infrastructure alone is not enough to assure agricultural and rural development and rapid increases in food production in developing countries. Yield enhancing production technology is of critical importance. Although opportunities for expansion of agricultural production into lands not currently under cultivation still exist in some countries, such opportunities are so limited that they would probably not be able to counter losses of current agricultural lands to alternative uses on a global level. Furthermore, attempts to expand agricultural production into new lands would, in most cases, require large investments in technology, tools and materials and would increase the risk of land degradation and deforestation. Thus, future increases in food production must come primarily from higher yields per unit of land rather than from land expansion.

Agricultural research has successfully developed yield-enhancing technology for the majority of crops grown in temperate zones and for several crops grown in tropical zones. The dramatic impact of agricultural research and modern technology on wheat and rice yields in Asia and Latin America since the mid- 1980s is well known. Less dramatic but significant yield gains have been obtained from research and technological change in other crops, particularly maize.

Natural Resource Management and Environmental Considerations

Research, technology development, incentives, and regulations are needed to prevent environmental degradation. These measures include appropriate water management policies, reduction of subsidies that encourage wasteful use of inputs, better definition of ownership and user rights to resources including land, education of farmers to encourage appropriate use of technology and resource conservation, and the provision of alternatives to resource-degrading inputs and techniques. Since poverty is a major source of degradation, poverty eradication is justified also on environmental grounds.

The recent surge in public and private concerns about negative environmental effects of economic growth and development may, if sustained, have important implications for agricultural development and future food production and consumption. Of particular concern of the need to avoid degradation of natural resources such as land and water, as well as deforestation, water contamination, and health risks associated with the use of chemicals. Since most of the current and potential resource degradation and environmental contamination result from situations in which those who cause and possibly benefit from degradation do not pay the costs neither the market nor the individual producers and consumers are likely to incorporate preventive measures into their behavior. Only when sufficient damage has been done to influence significantly current or future production costs will market and producer behavior change. The state is more likely to undertake preventive measures either through publicly funded research and technology development or through incentive policies and regulations. Extensive water logging, salination, and associated land degradation and productivity losses resulting from inappropriate water management are of particular concern in large parts of Asia.

No Time for Complacency

Population growth will outstrip growth in food production in sub-Saharan Africa for a long time to come unless more is done to accelerate agricultural growth. Between now and 2000, the production shortfall is estimated to increase to about 50 million tons of grain equivalent, up from the current level of about 14 million tons. The region will not have the necessary foreign exchange to import such large amounts of food. And African governments will not be able to count on enough food aid to make up the difference. If current trends continue, by the year 2020, Africa will have a food shortage of 250 million tons, which is more than 20 times the current food gap.

Poverty is expected to increase rapidly in the coming years. Sub-Saharan Africa's share of the world's poor is expected to increase from the current 19 per cent to about 28 per cent in 2000. Furthermore, the number of underweight children in expected to increase in the 1990s' in sub-Saharan Africa.

Asian demand for cereals is estimated to grow at an annual rate of 2.1 per cent between now and the year 2000, whereas food production is expected to grow at 1.9 per cent per year. Much of the production shortfall is likely to be dealt with through expanded imports and perhaps through expanded regional production in response to price increases.

In Latin America, by contrast, growth in food production is anticipated to exceed food demand growth: food production is estimated to grow by 3 per cent annually between 1990 and 2000, while food demand is estimated to grow by 2.5 per cent per year.

Large areas of land are rapidly being degraded and deforested. And the principal reasons for environmental degradation—poverty, high population growth, and limited access to appropriate agricultural technology—are not being dealt with effectively.

About 700 million people are food insecures for them the food crisis has arrived. For the 10-12 million preschool children who died in 1994 from hunger and diseases related to malnutrition, the food crisis came and went. One-third of the preschool children of the Third World are unable to grow to their full potential and face increased risk of death and disease.

Complacency is not in order. Clearly, Malthus underestimated the power of science to expand food production. The mass starvation that was predicted for Asia in the 1970s and 1980s did not occur because science was effectively put to work to expand crop yields. However, past yield increases came about people with foresight made appropriate decisions. The failure to expand investments in agricultural research and technology development during the 1980s and 1990s indicates that such foresight no longer prevails. Given the long lag time between investment in agricultural research and the resulting production increases, failure to invest today will show up in production shortfalls 10 to 20 years from now. The problems associated with environmental degradation will present themselves sooner. We must not wait until a global food crisis is upon us or until the last tree has fallen to make these investments.

25

Income Gap Widens

The gap in income among the people of the world has been widening. In 1960, according to United Nations statisticians, the richest 20 per cent of the world's people received 30 times more income than the poorest 20 per cent. By 1991, they were getting 61 times more. While the poorest one-fifth in 1960 received a meager 2.3 per cent of world income, by 1991 that revenue share had fallen to 1.4 per cent. The income share of the richest fifth, meanwhile, rose from 70 per cent to 85 per cent.

These disparities prevail both among countries and within them, and the large gap between individuals worldwide reflects the combination of both of those splits. Almost four-fifths of all people live in the developing world, where incomes are only fraction of those in industrial countries. In turn, within countries in both categories, gaps in income between citizens can be even wider.

The widest income gap reported within a country is in Botswana, where during the 1980s the richest 20 per cent of society received over 47 times more income than the poorest 20 per cent. Brazil was second, with a ratio of 32 to 1. In Guatemala and Panama, the ratio stood at 30 to 1.

The rapidly growing economies of East Asia have had income patterns similar to those of Western Europe and North America, with the richest one-fifth often earning 5 to 10 times more than the poorest fifth. In South Asia, India, Bangladesh, and Pakistan have had relatively even distributions of income, with the richest 20 per cent getting only four to five times more than the poorest quintile. Some countries that have had military conflicts apparently based in part on inequities among citizens, nevertheless have relatively even income distributions.

The split between countries and people can be seen in the marketplace. The value of luxury goods sales world-wide—high-fashion clothing and top-of-the-line autos, for example, exceeds the gross national products of two-thirds of the world's countries. The world's average income, roughly $4,000 a year, is well below the U.S. poverty line.

The poorest fifth of the world accounted for 0.9 per cent of world trade, 1.1 per cent of global domestic investment, 0.9 per cent of global domestic savings, and just 0.2 per cent of global commercial credit at the beginning of the 1990s. Each of those shares declined between 1960 and 1990.

These disparities are reflected in the consumption of many resources. At the start of this decade, industrial countries home to roughly a fifth of the world's population, accounted for about 86 per cent of the consumption of aluminum, and chemicals, 81 per cent of the paper, 80 per cent of the iron and steel, and three-quarters of the timber and energy. Since then, economic growth in developing countries has probably reduced these percentages. China's economy, for example, is more than 50 per cent larger now than it was in 1990, and developing countries have passed industrial ones in fertilizer consumption.

Uneven income distribution is shaping some of the most important trends in the world today. It raises crime rates, for example. And it drives migration. People have long responded to economic disparities by following a path from poor regions to richer ones, as tens of millions of workers chase higher wages and better opportunities. Some 1.6 million Asians and Middle Easterners were working in Kuwait and Saudi Arabia before they fled war in 1991, and at least 2.5 million Mexicans live in the United States.

The same is true within countries: rising disparities of income are adding to the growth of cities through rural-to-urban migration. Latin America, with some of the highest disparities of income among its citizens, is also the most urbanized region of the developing world-not entirely by coincidence. Since 1950, city dwellers there have risen from 42 per cent of the population to 73 per cent.

For many years, china had one of the most equal distributions of income in the world. But now that is changing, as incomes in its southern provinces and special economic zones soar while those in rural areas rise much more slowly. Also not coincidentally, the Chinese national Academy of Social Sciences forecasts that by 2010, half the population will live in cities, compared with 28 per cent today and only 10 per cent in the early 1980s.

In the early 1990s, developing world economies especially, in East Asia, have grown faster than the economies of the industrial countries. This has the potential to shrink disparities of income, if poorer countries continue to catch up. Yet even if the gaps among countries narrow, the gaps between people may not, because economic growth is distributed so uneventfly within nations. Despite the recent restoration of the economic growth in Latin America, no progress is expected in reducing poverty, which is even likely to increase slightly.

Meanwhile, in some regions almost no one has been getting richer. The per capita income of most sub Saharan African nations actually fell during the 1980's. In sub-Saharan Africa, the poorest geographic region, an estimated one-third of all a college graduates have left the continent. That loss of talented people, due in large part to poverty and a lack of opportunities in Africa, will make it even more difficult for the continent to advance.

The economic growth that has the potential to close income gaps among peoples in the developing world is instead becoming a splitting off, with some parts of societies joining the industrial world while others remain behind. Singapore, Hongkong, and Taiwan have begun to look like wealthy industrial countries, for example. Now parts of China are following, as are the wealthier segments of Latin American society and of Southeast Asian countries. This is good news for members of the middle-income countries and for the world. But it may do little to help the poorest fifth of humanity.

26

Trade and Labour Standards

Using the Wrong Instruments for the Right Cause

A moral value is a shared concern of humanity; hence its enforcement should be a cooperative task implemented for the benefit of humankind. Would we qualify recent approaches to the issues of trade and labour standards as non-inquisitory but shared and cooperatives ones? The purpose of this brief is to shed some light on this question.

In fact, nobody will deny any country the right to raise and fight for issues which are of moral concern for humanity, as they are supposed to benefit humankind. The issue of implementing and enforcing a core of labour standards one of these.

However, a problem remains: who has the negotiating power to raise and impose them? The key issue is that trade coercive attempts by some become inquisitorial as soon as they are backed by moral concerns which are supposed to be shared by all, while the same "all" lack the negotiating power to be, in turn, coercive if they so wish. In other words, trade-related coercion forcedly becomes "inquisition" when moral concerns are introduced into the functioning of an international trading system characterized by large imbalances in the negotiating power of the participating countries. Only a few governments have the negotiating leverage and strength to develop what we may qualify as "trade-related inquisitory practices".

The issue of trade and labour standards seems to have arisen when "uniform competition" has been regarded as a threat to employment and economic growth in some industrial countries.

However, without attaining a certain degree of international agreement and coherence as to whether and under what conditions – a given competitive advantage is, or is not, related to social or other conditions, and whether or not it may be considered as "unfair". With protectionist views in mind, such an approach may only be interpreted as "unbenign thinking" coming from "unfair competitiveness seekers".

If the motivation behind the introduction and further use of moral argumentation is to seek a justification for the possible use of trade measures as enforcement mechanisms to achieve certain goals. Particularly for harmonization of labour standards, one may wonder why the labour standards issue has not been linked to North-North trade in the current debate on the considerable variation in labour standards among developed countries. The motivation may well be that in the post-Uruguay Round era, when tariffs have been reduced substantially and "grey area measures" put under stricter control or even banned, we may be facing the possible revival of new forms of protectionism wearing "blue", "green" or "multicolour" masks. On the contrary, if the motivation behind the introduction of such moral labour rights argumentation reflects a real commitment by the international community to enforce labour standards, a door may be open for embarking, in the future, on a series on international initiatives, not necessarily under the trade umbrella.

It should also be stressed that the linkage between trade and labour standards has been seriously misinterpreted. Most analysts remain blind to the two-way character of the link between trade and labour standards. On the one hand, trade liberalization is to promote growth and development by promoting a more efficient allocation of resources and to ease the adoption and implementation of labour standards, as well as to promote job creation. On the other hand – and this is extremely important – raising labour standards – not keeping them low – should increasingly be seen as the real source of competitiveness and economic growth through, among others, increase in the quality of labour. There is a case for considering economic progress and the rise in labour standards as mutually reinforcing.

Turning to the low labour standards debate, much more empirical and analytical evidence is needed to assess the extent to which low labour standards are correlated to lower wages and labour costs. Although raising labour standards may not primarily be intended to maximize efficiency, it is becoming increasingly evident that efficiency and the future potential of the firm may not necessarily be maximized by keeping labour standards low. In this context, it is the development to human capital in LDCs which is a top priority, not because failing to respect labour standards in these economies threatens the welfare of the workers in the industrialized countries but, more simply, because it is the only strategy for enhancing the productivity of labour and ultimately increasing the people's standard of living. Thus, if raising labour standards and ensuring their effective implementation is important for economic progress, developed and developing countries, as well as workers and employers in each region, should adopt a cooperative, not confrontational, approach in order to deal with this urgent and pressing problem. For this very reason, adherence to trade sanctions would be a wrong approach. Trade is essential for enhancing workers' productivity because it ensures that a country's resources will be employed in the activities that it is best at. In turn, increased productivity is the key to development, higher labour standards and higher wages. Moreover, the issue of labour standards is of a moral nature: it has an undeniable development dimension which needs to be more clearly perceived but, as discussed, it is certainly not an issue to be dealt with through trade measures.

Labour standards should be dealt with in the WTO. The capacities of the ILO will prove invaluable for renewed multilateral effort to improve working conditions in developing countries. Its tripartite structure has proved to be the best suited to the tasks of conciliation and dialogue.

Improving the standards of living and labour standards for workers or eradicating child labour is what matters it is the right cause for humankind, a cause to fight for, through various approaches and by using all the mechanisms at our disposal in order to ensure its success. Thus, the issue is not one of trade and labour standards, but of labour standards and economic development, an issue of a human dignity and human rights nature. It is a universal issue, the solutions to which should be found by all nations, taking into account equity considerations. Each country should participate in the process, depending on its level of economic development. It is precisely because economic development, including trade, is positively correlated with the adoption and effective implementation of labour standards that solutions to low labour standards should not necessarily come from negative and coercive approaches. Solutions for universal problems must not only be efficiency-based but also equity-based. Therefore, the case is for cooperation rather than coercion and for applying positive instruments. As in the case of environmental issues, developing countries should be recipients of funds, technical cooperation and other similar forms of support when the implementation of labour standards involves an inequitable cost burden.

The initiative of "grouping" a core of existing conventions into a new global convention on core labour standards of universal value may, in view of its new and distant qualitative nature, generate strong support from the international community for its implementation. The ILO was created to promote workers' right, so the initiative could be launched under its aegis, in direct collaboration with other intergovernmental organizations dealing with social, trade and development issues in an interrelated manner. The support to this or other similar approaches by the international community, and in particular by those developed countries that have recently shown a special and strong interest in the reinforcement and implementation of labour rights in developing countries, will be a clear sign of their degree of sincerity and their willingness to share the social concerns of universal value which seem to be of paramount interest to most of their citizens.

27

Less Food Security in the South?

Combating hunger and poverty is the central point of bread for the world's mandate. In our view, that is not so much about the quantity of food produced in the world. On the one hand, it's about its fair distribution and, on the other, the access of poor people to chances of jobs. Put it another way, it's to do with access to purchasing power. In the case of agriculture, that is bound up with the question of how food is produced. Whether the technologies applied maximize employment or replace work with capital.

"Hunger through Surplus"

The question of production, employment and distribution is tied closely to the general conditions for development. It is certainly not exclusively external economic conditions which account for hunger and under-development. Structural deficits, political conditions and wrong policies in Third World countries have become increasingly clear. However, it can still be noted that global economic framework conditions remain enormously important for the development of agriculture in the Third World.

Twenty years ago, bread for the world publicly expounded the thesis "Hunger through surplus" and had to take much criticism for it—above all from agro-economists. But since then the contradiction between the ever-growing mountains of agricultural surpluses in the northern hemisphere and the increasing dependence on food imports of the South has become ever more apparent. Out of 120 poor developing countries, 107 today are net importers of food.

The North's surpluses of dairy products, grain, beef and sugar – which because of their production costs are exorbitantly expensive – thrust their way on to world market and destroy local supply systems (which are cheap because of subsidies), regional trade flows, and the sales possibilities of potential Third World agro-exporters. Thus, the surpluses contribute to the situation that in many developing countries a policy of neglecting local agriculture can be continued with impurity.

Initially, the promise to work on the yawning gap between hunger and surplus in the world was upfront on WTO agenda. But the pattern of explanation was well simplified. It said that surpluses arose only in those countries which supported their agriculture positively and, in fact, partly excessively. And that agricultural deficiencies in countries of the South were caused mainly by deprivation of resources and capital. However, the concept of not only reducing neglect of agriculture in the South but also its oversubsidising in the North to a sensible degree and thereby eliminating their distortions of world markets had a great intellectual attraction. At any rate, it promised more justice in agriculture.

Subsidies can Make Sense

To avoid misunderstandings, we have nothing against the support of agriculture in Europe. Above all not when it is done for social, ecological or agriculturally beneficial reasons.

On the contrary, agriculture's important role for food security, the sustainable handling of natural resources, the settlement of rural areas, and the social function of family farms justify a special position for it is economic life, including protection and support.

But that must not be carried so far that surpluses are produced with the help of dubious production methods and then dumped on the world market at markedly less than cost price, causing incalculable damage in the poor countries. On the other hand, purposeful promotion of rural development is a prerequisite and model for greater self-sufficiency worldwide, especially in Third World countries.

Complementary Functions of World Markets

The poor countries of the South have no alternative than to become self-sufficient in food. The world markets can at best assume complementary functions. The countries would take indeterminable risks if they integrated themselves completely in the world markets, and thereby wanted to make themselves dependent upon global agro-markets. These are and will remain extremely unreliable factors that are conditioned by enormous fluctuations in prices and quantities, the powerful, and in many cases obscure, influences of multinational concerns, the manifold political interventions in the agricultural scene in most countries, and the danger of social and ecological dumping.

But when we now look at the results of the WTO, we are disappointed. The development question and the balancing of hunger and surplus are finally no longer on the agenda. Programmes to increase food production in the poor countries were not the priority of the negotiations. The liberalization of agro-policies in the developing countries would have meant making the disadvantaging of their farmers the subject of international negotiations. That did not happen.

On the contrary, the concepts developed with an eye on the reform of agricultural policy in the North, which target the reduction of the support level, are to be transferred to the South without questions. To be sure, there are a whole number of exemptions for the poorest developing countries. But the WTO results have also set the trend there, namely, the dismantlement of subsides. We cannot understand how such a thing can be demanded as a policy programme, especially for Africa. Support for African agriculture is largely absent, i.e. there is absolutely nothing to dismantle. That's why many international conferences repeatedly emphasize the need for these countries to achieve a greater degree of self-sufficiency in food by stronger support of their agriculture.

Agro-Dumping

Certainly, some changes have been made in the North's agro policy system which will also have positive impacts on world agricultural markets. However, also here we must express our disappointment. Agricultural dumping will continue. The only difference will be the new policy instrument of direct transfer of income instead of subsidized grain prices. The opening of markets in future will hardly go beyond the current preference conditions.

The entire set of W.T.O. agreements, however, bears the imprint of the two agricultural superpowers, the USA and the European Union, which make mutual concessions and coordinate their agricultural policies. But one hears nothing about the target of freeing the world agricultural market from unnecessary distortions and ensuring justice. The intention of the agro-superpowers was solely to defend their global market shares.

The development aid agencies cannot close their eyes to these problems. On the contrary, in future they must make very much greater efforts in suggesting better goals, programmes and instruments which are capable of forming a policy that can then be included in the agenda of the next rounds of negotiations. We may perhaps have slept a bit through the past WTO talks. Therefore it is even more important that we get very much more involved from now on.

28

Safety First!

Throughout history, the violent side of nature has manifested itself in destructive phenomena such as floods, volcanic eruptions, severe storms, wildfires, earthquakes and tidal waves. Disasters and risks are part of our life, and they will continue to threaten, kill and destroy. "Zero risk" is out of reach in the contemporary world. The basic problem is how to prevent hazards from causing increasingly large-scale disasters. The answer is that the disastrous effects of natural phenomena will only be eliminated, reduced or stabilized when people decide to make cities, settlements, infrastructures and houses safer.

A Culture of Prevention

For people are the agents of disaster. "It's not the bullet that kills, it's the hole". Earthquacks and windstorms do not kill; the collapse of houses and buildings from shaking is the main cause of death. Natural hazards themselves are not on the increase; nor are they likely to be in the future. It is the frequency of natural disasters that is expected to grow, as well as their complexity, scope, gravity and destructive capacity. There will be an increase in multiple or synergistic-type disasters causing society-wide impacts. In a time of globalization, largescale damage caused by an earthquake in a world financial centre is bound to have an effect even on the economies of faraway countries.

Natural disasters are not always entirely "natural". On the one hand, natural forces are at work on planet whose environment is being altered day after day by humankind: floods are made moreover by deforestation, global warming more preoccupying by the unchecked emission of greenhouse gases. On the other, natural disasters will increasingly generate or magnify concurrent technological disasters. Floods can devastate chemical complexes, earthquakes can affect critical plants.

The good news is that disaster reduction is both possible and feasible. While we cannot prevent an earthquake or a windstorm from occurring, or a volcano from erupting, we can use the scientific knowledge and technical know-how that we already have in order to increase the earthquake—and wind-resistance of houses and bridges, and to issue and disseminate early warnings of volcanic eruptions and organize proper community response to such warnings. The extent to which society puts this knowledge to effective use depends upon its social, cultural, political, economic, and even religious specificities.

Informing the Public

Disaster prevention and preparedness start with improving our understanding of risks by assessing the distribution in time and space and the intensity of the natural phenomena involved and the exposure of people and structures to them. On the basis of this assessment, protective measures may be taken such as land-use restrictions, adequate construction measures and wise environmental management. Detection and warning systems may be installed, and contingency and emergency plans be set up. One permanent measure of paramount importance is the education and information of the public.

A number of cases show that loss of life, injuries and physical damage can all be significantly reduced by the application of better warning and disaster prevention measures. Because of inadequate use of, and response to, warning, more than 300,000 people died in Bangladesh due to a cyclone in November 1970. In May 1985, better prediction and proper response to warning of a cyclone of the same intensity kept the death toll below 10,000. Similarly, appropriate warning and evacuation saved the people in India in various cyclones.

Disaster prevention measures cost much less than relief and reconstruction expenditure following a disaster, yet many decision-makers tend to focus on relief and to treat disaster situations in an ad hoc way when they are presented with them. Today most typical strategies are crisis-oriented. Furthermore, information about natural hazards and disaster reduction techniques is not well disseminated, and planners, project managers and communities do not integrate hazard management into development planning. Resources spent on relief and recovery continue to account for 96 per cent of all resources spent on disaster-related activities annually, leaving a pitiful 4 per cent for disaster prevention. It is high time to make a shift in emphasis from post-disaster reaction to pre-disaster action.

29

Development: The Third Way

While great claims are being made for the increasingly more efficient and effective technologies perfected to serve development of the people in this scientific age, huge problems are threatening the globe. The problems are mass poverty and hunger, underdevelopment, waste, unemployment, resource scarcity, environmental destruction and armed conflict.

In finding answers to the prevailing problems we must be clear about the meaning and purpose of development. The first glaring mistake made is that development is interpreted as development of the economy and not as the total development of society. When economic development is made the supreme goal, most of the other vital aspects get ignored, namely, development of the political system, community, social cohesion, the ecology, culture and values. Development and stability go hand in hand while poverty and chaos constitute the antithetical twin.

Appropriate Development

The key elements in the conception of appropriate development consist of: first, aiming at sufficiently comfortable material living standards and not affluent standards of the rich as in prosperous nations. Second, development must not be confused with GNP growth. Mere increase of economic activity must not be pursued exclusively at the cost of articles that are urgently needed by the poor majority to maintain them at a reasonable level of material living. Third, in the villages, we must produce articles as are needed by the villagers. Fourth grassroots and participatory development is essential so that the local people identify and solve their local problems. Fifth, instead of capital and energy intensive high technology, use labour-intensive technology. Instead of heavy industrialization, promote medium scale industries and technologies. And, sixth, instead of preoccupation with a high GNP growth rate, focus on the development of communities and of rural bodies and take care to conserve the local ecosystems. The main purpose should be to meet the primary needs of ordinary people and promotion of their productive resources such as land.

In developing countries like India where billions of poor people remain condemned by conventional economic development strategies and theories, it is vital to introduce appropriate development measures to remove deprivation and ensure the necessity of modest living standards.

The world has witnessed the operation of the two systems, namely, the capitalist and the socialist one that have obtained in different countries. Although both the systems have underlined the welfare of all as the basic goal, both have left a legacy of waste, hunger and gross human inequality. Some 1000 million people do not get enough to eat including some 20 million in the USA.

Third World Way of Development

The iniquitous situation in the present-day world has sparked off fierce controversies among the conventional economists and the new radical economists who champion a third way, as the alternative way to serve the primary goal of all humanity to have sufficient means to lead a comfortable peaceful life.

In order to achieve prosperity, conventional economists have emphasized the production of bigger cake on the assumption that everyone will get a slice of it. They also argue that a "tide will lift all the boats". Both these assumptions have proved false in that the poor have neither the slice of cake nor has their boat been lifted. Third way system lays stress on highly localized, less cash-reliant and simply structured set-up. It should not be dependent on transport of goods, but concentrate on more local production to meet local needs with a role for barter and free exchange. He urges a radical re-think of conventional economics.

Is this Stepping Backwards?

The most common criticism levelled against the Third Way is that it will arrest the progress made hitherto and that it might mean a return to a 'primitive' way of life. There should be no fear no this score because the Third Way aims at the reduction in the use of resources and therefore, of excessive production and consumption. It does not in any sense mean stepping backwards to a lower level of the quality of life. Nor is the alternative way intended to destroy capitalism or socialism. The conception of the alternate way is to promote economic growth compatible with capitalism and socialism. The ground idea is to promote selflessness, mutual concern and social responsibility. This will replace selfish, competitive and avaricious attitudes as have developed in the conventional economic order of today.

NGO's Resolution at the Rio Conference

That there is increasing awareness of the threat posed by the growing power of multinational corporations was articulated by the international NGO forum in its Declaration resolved on 12 June 1992 at the UN Conference of Environment and Development in Rio de Janeiro. The Declaration states that "the Bretton Woods institutions have served the major instruments by which the destruction policies have been imposed on the world" and calls upon "the world's people to protect their economic, social, cultural and environmental interests against the growing power of transnational capital". The declaration further avers "we recognize the central place of spiritual values and spiritual development... and values of simplicity, love, peace, and reverence for life.

After the failure of the socialist system over four decades to achieve prosperity for all, the Indian Government switched over to the global market economy and is steaming ahead with added liberalization measures to attract foreign investment and the multinational corporations (MNCs). Some adverse effects of this are already visible: for example, majority financial equity granted to MNCs and the emergence of foreign subsidiaries with cent per cent financial equity; the introduction of pizza and Kentucky Fried Chicken which has been detested by the people in Karnataka. The farmers have also revolted because their rights to produce and sell seeds have been wrested by foreign MNCs who have acquired patent rights over certain Indian crop seeds. In this scenario, the Third Way has much to commend itself to the Government. The Third Way has the air of the Gandhian model of economy and production which emphasizes production by people for their own needs and preference for small and medium sized industry. The same paradigm was championed by the renowned economist Schumacher when he said, "Small is beautiful". India should take good care against the present-day headlong drive for the entry of foreign capital and foreign heavy industries.

30

Democracy and Poverty: Are They Interlinked?

Democracy assistance and poverty reduction are rightly becoming two focal—and related—issues for development assistance. Increasingly, many organizations, including intergovernmental, national and civil society, are focusing their work on these two areas. Furthermore, the relationship between these two issues is complex and ever changing. There is thus a need to develop methodologies for linking democracy assistance and poverty reduction at both the policy and programme levels. International IDEA (Institute for Democracy and Electoral Assistance) in cooperation with the World Bank and the United Nations Development Programme, is developing concrete strategies that address these two objectives in a mutually reinforcing way. Through an overall situation analysis followed by regional meetings in sub-Saharan Africa, South Asia, Latin America, the Caucasus and the Arab region, the Institute has marshalled evidence of some of the key problems that affect democracy consolidation and poverty reduction in these countries:

- Corruption and its undermining effect on popular confidence in public institutions.
- Continuing economic instability coupled with the lack of strategies for addressing the twin challenges of poverty and increasing popular participation in its alleviation.
- The extremely limited nature of citizens' influence on overall policy and decision-making process despite the spread of formal democratic institutions.

- A trend in many post-communist states towards viewing growing poverty as a direct consequence of a transition to democracy.

In short, the evidence is not very encouraging for the prospects for democracy consolidation and poverty reduction. The critical step International IDEA advocates is the development of an approach that not only seeks to put democracy assistance and poverty reduction on top of the development assistance agenda, but also to encourage all involved to treat them as twin elements of an integrated programme of action.

Through a focus on accountable governance, promotion and protection of citizenship and rights and increased popular participation, International IDEA believes that both democracy and poverty reduction can be addressed simultaneously. Policy recommendations are being developed and will be shared in the course of this year with governments, international organizations and civil society bodies.

International IDEA believes that democracy promotion can be used as a tool for fulfilling a variety of objectives. Democracy matters because it protects human rights and preserves human dignity. But democracy also matters because it helps to address some of the most critical challenges facing states today: peace, development, economic growth and stability.

Democracy does not guarantee any one of these, but increasingly it seems to be a precondition for them in the long term. Thus, advocating democracy goes beyond being a moral issue; it becomes *fundamental* to advancing the well-being of people and the stability of states. International IDEA will continue to explore the link between democracy and the major issues facing society today—and continue to argue the case for democracy.

31

Tobin Tax

James Tobin, a Nobel laureate in economics, dreamt up his idea for a 'Tobin tax' in the 1960s when the gold exchange standard was still in force. Before long the US closed the gold window' at the Federal Reserve, giving rise to the current regime of floating currencies. Tobin argued that speculative flows of 'hot money' would put unwanted pressure on fixed exchange rates not reflecting economic fundamentals, and would upset macro-economic management he thought that if purchases and sales of foreign currency were subject to a small tax (a fraction of 1 per cent of the value of each transaction), long-term investment and current account operations would not be significantly affected, but currency speculators conducting large volumes of transactions adding upto a high value would be deterred.

The Tobin tax has never been implemented and was largely forgotten until in the mid-1990s. The late Mahbut-ul-Haq and his fellow authors of the United Nations Development Programme's annual Human Development Report then revived the idea with a view not so much to calming fever in the international capital market as raising extra resources to finance development. Now, War on Want is echoing this call in its report Costing the Casino: The Real Impact of Currency Speculation in the 1990s.

Currency speculation plays a central role in global instability which in turn brings increased poverty and instability to developing countries, argues War on Want. Its report shows, for example, that the delayed effects of the South East Asia economic crisis in 1997/98 continue to hit the poor.

At a time of declining official flows, developing countries rely more and more on private capital flows to finance their growth. Their dependence on volatile, short-term investment exposes their fragile economies to sudden changes in financial markets. The frequency of financial crises around the world in itself urges measures to stabilise this volatility.

A Tobin tax could help deter speculation by making currency trading more costly, in particular it would act to deter short-term transactions of large sums between countries more expensive.

Another advantage of the tax, says War on Want, is its revenue-raising potential—particularly in view of the 'current diminishing scale of aid budgets'. The organisation believes the threat of further economic crises, as in South East Asia, makes the adoption—or at least consideration – of a Tobin tax 'a matter of some urgency'.

However its critics say the Tobin tax raises a number of thorny questions: Does it make economic sense? The consensus view of economists is that markets, including capital markets, work best if information is good, transaction costs are minimised and distractions removed. The Tobin tax would introduce an, albeit small, transaction cost and distraction. True capital markets are subject to bubbles and panics. But taxes are not the way to deal with these. Instead, there needs to be better supervision of markets and institutions, with occasional official (or IMF) intervention to preserve stability?

Would the Tobin tax work? If applied, it would almost certainly be subject to massive avoidance as tax-liable transactions were relocated in financial centres whose authorities had not agreed to implement the tax. There are no internationally-recognised means of obliging countries to raise a tax. Furthermore, there is a very rapidly increasing volume of transactions in foreign exchange derivatives—futures, options and swaps. It is not clear if these would be taxed also.

Would the tax raise significant resources for development? Very large sums could theoretically be raised if foreign exchange markets all applied the tax and if there were no avoidance, and if the governments which collected the tax were willing to devote the proceeds to aid. But governments have no overriding reason to hypothecate the revenue from any particular source to aid (why not devote the proceeds of airport departure tax to aid?)

According to its critics, the Tobin Tax was not a very good idea for achieving its original purpose of calming speculations, for both theoretical and practical reasons. Nor is it a good idea as a source of more aid for the same practical reasons (is it workable?). Even if these were overcome it would require the assent and collaboration of the very governments whose current decisions to provide aid (or not) have hitherto determined the magnitude of aid flows.

32

Resistance to Change

Why Poverty Reduction Programmes Did Not Work

Poverty reduction as an overall objective of the global development industry is not new. The only problem is that so far it has not really worked. Despite several decades of economic growth and huge development aid disbursements, the number of countries the United Nations calls "least developed" (those with a per capita income of less than $900US a year) has in fact nearly doubled since 1971, from 25 to 49. In the last decade (1990-2000) and despite all development efforts – not even one country was able to graduate from this group to a higher income level, may be with the exception of Botswana.

Meanwhile, poverty reduction has generated its own history. This programme has covered a wide range of approaches starting from the World Bank's small-farmers-strategies in the 1970's via the costly structural adjustment policies of the 1980's to the recent poverty reduction strategies of the 1990's. Once more, the next development decade (2000-2010) has written "Attacking Poverty" on its banner. It seems that something must have gone wrong along the way. What (bitter?) lessons have been learnt from previous experience? Have they been factored into the new set of policies? Were there possibly some fundamental flaws which were overlooked, and can better results be expected during the next period? Or do the many failures and disappointments demonstrate that there is some systemic "resistance to change" by those in power in the least developed countries and perhaps also by the poor themselves?

1. **What can the rural poor really expect from poverty reduction programmes?**

In India, for example, 70 per cent of the people still earn their livelihood in the agricultural sector; most of the poor among them live in a kind of rural subsistence economy. People who live in a subsistence economy are naturally conservative. They are busy securing their survival and are very reluctant to take risks. Their living standard is measured in amounts of rice harvested; their wealth is measured in numbers of livestock. Within this simple framework, poor peasants behave very rationally. For example, a shift from food crops to cash crops, such as from rice to coffee or tapioca, would immediately endanger their subsistence in case of failure. Furthermore, the poor do not have the knowledge and skills to change their crops quickly in response to market demands. Moving from a subsistence economy to a commodity economy is therefore a big step for small farmers.

However, poor people are always happy to receive handouts from the government like fertilizer, seeds, medicine or blankets. Roads, bridges and schools are also very welcome. Who would refuse a gift? From their point of view, it is the responsibility of the Government to distribute goods and services in form of aid programmes as a way to share some of the prosperity of the city people with them. Nevertheless, as they see no direct and immediate benefit for themselves, they tend to take a rather passive attitude to change. Development workers have often complained about this common apathy and about the lack of will among the poor themselves to improve their situation. In the final analysis, rural development is more a problem of providing the right economic incentives for change than of overcoming traditional thinking and a conservative attitude.

2. **What kind of incentives are necessary to achieve increased production in the countryside?**

In most poor countries the key to rural development is the problem of land ownership rights and of legal security. As long as people do not own the land that they cultivate, they are not interested in making any investments, be they in the form of labour or capital. Once a farmer has an ownership title and considers the land as his own, he will refrain from over-using the soil but shift crops and plant new trees. Moreover, he can then use his land as collateral for credits or even sell it and buy land somewhere else.

In addition to clear and irrevocable ownership rights, the rule of law is another crucial factor for development. People must feel safe from abuse of power by local elites and corrupt government officials. They must be able to enforce their basic rights in an impartial court of law. Furthermore, they must be safe from land expropriation without adequate compensation and from resettlement against their will. In other words, it is primarily their very stake holdership in the rural economy that will motivate them to increase their production. Of course, the other necessary incentives are access to markets, a fair price for their products and the availability of goods and services.

3. **Poverty reduction programmes, if not accompanied by parallel institutional reforms, run the risk of crating a modern version of the cargo cult.**

Cargo cults spread during World War II in the highlands of Papua New Guinea at a time when several US cargo planes loaded with food supplies crashed into the hills. Suddenly, the native people could enjoy an abundant amount of goods, which literally fell down on them like a "gift from heaven". In the hope of attracting some more of these "silvery birds", the local hilltribes constructed primitive models of airplanes, sat around them in a circle, and prayed that more "cargo" would drop on their territory. As this happened in some areas (albeit as a result of the air battle between Japan and the USA), it strengthened the belief in the cargo cult as some magical way to overcome poverty – at least for a short time.

There is a high risk that aid programmes under the banner of poverty reduction will create new "cargo cults" in the 49 least developed countries if they continue to carry out their "business as usual" and do not put strong emphasis on the rule of law and civil rights. Unfortunately, the setting up of reliable legal and social institutions in poor countries (which often seems to be the "software" of the development industry accompanying disbursements) is, in fact, as decades of experience have shown, the hard part of the process. But it is also indispensable for achieving any tangible results.

Why have there been until now only modest results in the areas of land reform, rule of law and the guarantee of basic civil rights? Why have people's participation and people's ownership as a strategy hardly taken root at all in the least developed countries? The answer must be sought in the role of powerful local groups and their vested interests, who obviously benefit from the prevailing status quo and a loose legal environment. A cargo cult promises bounty for all recipients; poverty reduction, however, means changing the rural power structure, too.

Conclusion

To insist on the rule of law, on people's participation in the development process, and on transparency and accountability, is again nothing new. Good political and administrative institutions go hand in hand with economic growth. The potential of economic development is quite limited if it works in a framework of social undevelopment and official indifference. Again the question is, who has so little been achieved in this field during previous decades? Was it the wrong medicine and why were the poor results of the aid programmes so carefully ignored by the international donor community?

Looking at the political systems of the 49 least developed countries, it is obvious that most of these countries are "more democratic in principle than in practice" Many of them are ruled by military or civil authoritarian regimes which are more used to giving orders than to listening to the grievances of the poor. Other governments, such as India, are "genuinely democratic at most levels but have historically found it difficult that political accountability reaches all levels of decision making, particularly for the poor".

To sum up, it seems that resistance to change is equally shared by the cumbersome and often incompetent bureaucracies of the poor countries and the equally cumbersome international donor community, which has so far conveniently kept the call for more rural democracy and people's rights on the backburner. The major reason for the reluctance of the donor community to pursue the battle for the rule of law and the fight against endemic corruption was to avoid massive political confrontation with the receiver countries.

Would it not have been better to create proper incentives for the performance of poor countries, namely by halting loans to nations that do not manage their economies and their reform commitments effectively and increasing financial and technical support to those that do? The next decade will show how determined both local governments and donors are to tackle these problems for the sake of a better future.

33

The Role of Financial Markets and the IMF

The Genesis of Asia's Financial Crisis

It is obvious that the major causes of Asia's financial crisis were rooted in the affected countries. The evils were excessive foreign borrowing, poor supervision of banks, and overvaluation of national currencies. Apart from those policy failures, however, there were external factors. These were, besides the instability of the international finance markets, above all the wrong reaction of the International Monetary Fund, which aggravated the crisis rather than counteracting it. The role of the IMF must be fundamentally redefined.

The Asian crisis marks for the time being the end of the Southeast and East Asian economic miracle of the last four decades. Engulfed by the malaise of the worst-hit countries, Thailand, Indonesia and South Korea, the entire region is suffering economic weakness and in part even a decline in economic performance. The question is whether the Asian crisis could have been avoided or whether wrong economic policy decisions gave it a kick-start.

The Causes of the Crisis

Analysis of the causes of the Asian crisis has to date been marked by an astonishing one-sidedness. Upfront, explanations emphasise the internal causes, particularly the private sector's excessive foreign borrowing. But two other factors played central, if not decisive, roles in the spread of the crisis. One was the great volatility of the international capital markets, the other the inappropriate policies of the IMF.

This article analyses these three major causes of the crisis. Another question is how in future one can prevent manageable economic problems getting out of hand and developing into a crisis that threatens more than the economic stability of single countries.

But a closer look shows that the crisis, which broke out in mid-1997, has assumed such an unforeseen dimension that individual corrective measures of economic policy can no longer cope with it. A huge structural economic crisis has arisen. What developments led to it?

Foreign Borrowing

Let us look first at the high private sector borrowing abroad. With hindsight, it is easy to pass judgement on it as having been wrong and dangerous. But that does not mean that private foreign borrowing is harmful in general. If it finances profitable investments, it is merely the use of foreign savings when domestic savings are too low. IMF reports also reflected this assessment.

We know from experience of earlier debt crisis that a high foreign indebtedness by the private sector is latently, but not generally, risky. The Asian crisis also has reconfirmed that international finance markets differentiate between individual private debtors and the credit-worthiness of national economies only in the case of a few countries. In smaller countries, including OECD member South Korea, difficulties in servicing individual loans lead to investors getting out of these markets. This is the real danger of large-scale private foreign debts.

That makes foreign loans much more expensive than domestic credits. It the national interest rate is higher than that of international market, cash deposit requirement often can be an adequate incentive to borrow at home or take out a longer-term loan. Both alternatives lead to greater stability of the domestic finance system, as the risk of withdrawal of capital at short notice, as in the case of the Asian crisis countries, is much reduced.

Instability of the Finance Markets

Beyond the cash deposit requirement, however, further restructuring must be done in the finance sectors of the Asian crisis economies and threshold and developing countries. Frequently, there are demands for greater transparency. Improved banking supervision and, especially in the case of South Korea, a broader diversification of shareholding. These steps are important, but they will not prevent the next crisis because they do not put an end to the instability of the international finance markets.

The Asian crisis has also reconfirmed that highly mobile capital can produce instability. It must seem dubious when there are calls (particularly by the IMF and US politicians involved in financial matters) for the pushing through of even more capital mobility as a consequence of the crisis. But even liberal economists are now increasingly questioning the ideal of a world without restrictions on the free movement of capital linked with an enhanced IMF mandate.

IMF Intervention

The IMF's policy intensified the volatility of the international capital flows, that is, their wide fluctuation margin when American and European fund managers began to pull their capital out of Asia's crisis countries. Asian debtors fell into arrears in servicing their liabilities, and the currencies came under heavy pressure, the IMF ordered a drastic cure. This was aimed at lowering inflation rates and reducing government budget deficits. But as there were no critical inflation rates, and government budgets were, in fact, in surplus, this policy was more than odd. A quick look at the economic development in the Asian crisis countries illustrates the relatively positive situation of the three economies before the crisis broke out.

What Measures to Stabilise Currencies?

The IMF policy also manifests deficiencies beyond the measures it ordered. I shall not discuss here the question of whether it is wise and appropriate for the IMF to clamp measures and restrictions of private sector debtors and their governments while the creditors side can emerge from the crisis largely without loss.

But one must ask whether the IMF's measures to stabilise the currencies were suitable in an acute-crisis. They are mainly measures with individual impacts, giving priority to increasing real interest rates. The aim is to regain the trust of international finance markets and stimulate fresh loan flows to these countries. The long-term objective is to restore currency stability.

This policy, however, has two weaknesses. On the one hand, it sets indebted companies under more pressure as they must not only pay more for their foreign currency loans due to devaluation, but are also faced with higher domestic interest. On the other hand, while it is true that a high interest rate policy in countries with stable economies can have a positive impact in attracting capital, that is not so in countries suffering from an acute economic crisis, as Indonesia shows. Recovery of the exchange rate to a realistic level that would enable the companies to service their debts has not happened there.

The Future of the IMF

Having reflected upon the dubious results of the IMF policy, one is bound to ask what role the organisation should play in future. Should its task be to have a stabilising effect in a crisis, or should the IMF be an instrument to assert certain economic policy concepts?

The IMF itself has defined its current function very one-sidedly, focusing on its service for the international finance markets. According to a self-assessment in an internal IMF document, it sees itself performing a dual function as a 'confidental economic adviser' and as the 'watchdog for the international financial markets'. But the IMF does not by a long way give the attention they deserve to the interests of the 350 million people in the countries under its wing.

Criticism in the West

Criticism of the IMF is growing not only in the Asian crisis countries but also especially in the USA and more and more in Europe. Conservative American politicians such as the former US Secretary of State Charles Shultz have described the IMF as ineffective, unnecessary and obsolete. They have also proposed that the IMF be abolished at some time after the Asian crisis has been overcome. The IMF's current policy is also coming in for heavy criticism in the academic debate on the crisis, which is demanding a redefinition of the organisation's mandate.

Other possibilities are conceivable beyond the radical option of abolishing or privatising the IMF. The IMF should in any case be required to tackle the specific situations in the crisis countries with greater awareness of the affected countries. In the course of the Asian crisis, the option of creating a regional fund was also discussed, but the Western G7 countries and the IMF emphatically rejected it. One should, however, consider whether regional institutions could not in fact work more efficiently than an authority based in Washington with competence for the entire world.

A regional structure with several monetary funds could facilitate the overcoming of crisis situations, although only if a global structure were to be maintained alongside the regional components. This global body, a kind of world monetary council, would be composed of representatives of the US and European central banks and the regional funds. Besides taking over the IMF's current tasks, such an international regime could also deal with stabilising exchange rates between the industrialised nations and in particular with the development of a target zones system between the dollar and the Euro. The winners in a less unstable international finance system would be the developing and threshold countries that at present still need to take the questionable medicine prescribed by the IMF.

34

The Nature and Causes of Drug Addiction

Man has been experimenting for thousands of years with a variety of naturally occurring substances that act on his nervous tissues: alcohol to intoxicate a weary mind, belladonna to calm an angry intestine or to poison an adversary, opium to overcome worry and strain. The relief of pain, in particular, in an age-old aim of mankind, and various narcotic and sleep-producing agents were probably used by primitive man. But for many men there is another kind of pain—the pain of being—and from time immemorial some men have been trying to expand their vision, enhance their appreciation of their world, change their mood, alter their inner existence, or stupefy their awareness with such drugs as alcohol, opium, and cannabis.

Drugs, chemical substances that affect the functions of living things, are used in treating, preventing, and diagnosing diseases. The most important source of drugs today is chemical synthesis. The increase in the manufacture of drugs has resulted in the development of 25,000 or more drugs and drug products. Many drugs are potentially dangerous chemicals that can cause serious, sometimes fatal, poisoning if used incorrectly: governments of various countries, therefore, have established certain legal requirements concerning drug use.

Uses: The purpose of the use of drugs is to cure disease or correct a disorder. Chemotherapeutic drugs, such as the antibiotics, the sulfa drugs, and the antimalarial drugs, flight infection by acting directly on disease-causing invading organisms, either immobilizing or killing them. Some chemotherapeutic drugs are also used to suppress or prevent infection.

Drug Toxicity

No drug is free of toxic effects. This factor is what ultimately limits the usefulness of drugs. Some of the untoward effects of drugs are trivial and can be readily tolerated. Other, however, are serious and may even be fatal. Some toxic effects of drugs are merely extensions of the drug's therapeutic effects.

This is why drugs never should be taken except under the guidance of a physician. A physician is aware of the potential hazards of a drug and is prepared to act promptly if toxicity occurs. Furthermore, the physician is aware that many of the toxic effects produced by drugs are unexpected, bizarre, and often not clearly related to the taking of a drug.

Many people suffer from drug allergy—one of the most serious problems of pharmacology. Penicillin, for example, is an extremely safe drug for most people, but it produces hypersensitivity reactions in about 15 per cent of the population. In some cases the reaction is so serious that it is necessary to forbid the future use of penicillin because of the risk of death. Drug allergy takes many different forms; skin reactions varying from a mild rash to severe dermatitis.

Drug Addiction and Abuse: It is very likely that every society has had mood-changing drug and that there have always been individuals who used them in ways that were not socially approved. In this sense, drug abuse, the socially nonsanctioned use of a drug, is universal and so old as history. Which behaviors are called drug abuse varies from culture to culture and from time to time within the same culture. Since laws do not always correspond to prevalent social attitudes, there may be times when users of an illegal drug are not considered to be drug abusers. From a pharmacological viewpoint, attitudes toward drugs are often inconsistent or irrational. Some drugs may be totally outlawed, while others with similar actions are made generally available and may be self-administered with full social approval.

The repeated use of some drugs can lead to a dependence on the drug in which the effects of the drug or the conditions associated with its use are felt by the users to be necessary for their well-being. Dependence may vary in intensity from a mild inclination to a strong craving or compulsion to use the drug. Severe dependence may result in a type of behaviour as also known as compulsive drug use, and since a severe dependence on any self-administered drug is generally not socially approved, the term is usually synonymous with compulsive drug abuse. One obvious exception is the use of tobacco, where social acceptance is so complete that even heavy compulsive use, which is damaging to the user's health, is commonly not considered to be drug abuse.

The term drug addiction has been defined in many ways, but in this article it is used to mean a behavioral pattern of compulsive drug use characterized by an overwhelming involvement with the procurement and use of the drug and the high tendency of the user to relapse to drug use after a period of abstinence. It is synonymous with intensive or severe drug dependence. Contrary to popular belief, drug addiction is not the same as physical dependence on a drug. Physical dependence is a physiological or biochemical condition produced by the administration of a drug to the extent that a characteristic pattern of signs and symptoms appears when the drug is withdrawn and disappears when the drug is administered again. Physical dependence can be produced by a wide variety of drugs that are used in everyday medical practice. Some drugs that produce physical dependence are not pleasant to take and are neither abused nor used compulsively. Also, not all withdrawal symptoms are associated with a craving for the drug that produced the physical dependence.

The Nature and Causes of Drug Addiction

If opium were the only drug of abuse, and the only kind of abuse where one of habitual, compulsive use, discussion of addiction might be a simple matter. But opium is not the only drug of abuse, and there are probably as many kinds of abuse as there are drugs to abuse, or indeed, as may be there are persons who abuse, Various substances are used in so many different ways by so many different or one definition could possibly embrace all the medical, psychiatric, psychological, sociological, cultural, economic, religious, ethical, and legal considerations that have an important bearing on addiction. Prejudice and ignorance have led to the labeling of all use of nonsanctioned drugs as addiction and of all drugs, when misused, as narcotics. The continued practice of treating addiction as a single entity is dictated by custom and law, not by the facts of addiction.

Many substances are capable of acting on biological systems, and whether a particular substance comes to be considered a drug depends, in large measure, upon whether it is capable of eliciting a "drug like" effect that is valued by the user. There is nothing intrinsic to the substances themselves that sets one active substance is imparted to it by use. Caffeine, nicotine, and alcohol are clearly drugs, and the habitual, excessive use of coffee, if not addiction. The same could be extended to cover tea, chocolates, or powdered sugar, if society wished to use and consider them that way. The task of defining addiction then is the task of being able to distinguish between opium and powdered sugar while at the same time being able to embrace the fact that both can be subject to abuse. This requires a frame of reference that recognizes that almost any substance can be considered a drug, that almost any drug is capable of abuse, that one kind of abuse may differ appreciably from another kind of abuse, and that the effect valued by the user will differ from one individual to the next for a particular drug, or from one drug to the next drug for a particular individual. This kind of reference would still leave unanswered various questions of availability, public sanction, and one kind of effect rather than another at a particular moment in history, but it does at least acknowledge that drug addiction is not a unitary condition.

Effects on the Mind and Body. The effects of drugs are similar in many ways to those of alcohol. Low doses usually produce relaxation and decreased anxiety; higher doses produce drowsiness. Even if people can stay awake, they may appear confused and show poor judgement and loss of emotional control. Slurred speech, a staggering gait, muscular incoordination and nystagmus (rapid involuntary eye movements) are also characteristic effects. Although alcohol and the sedative-hypnotics are all depressants of the nervous system, low or moderate doses can produce an effect that resembles stimulation. The individual may become euphoric and more active, and show a decrease in inhibitions. Very high doses produce coma and death due to respiratory failure.

Drugs in Psychiatry. Drugs that are used either alone or in conjunction with psychotherapy to treat psychiatric illness. The medical treatment of psychiatric illness is based on a firm conviction that the patient's behavior is in fact a symptom of an illness and not simply a variant of acceptable behavior in society. The study of drug effects on mental processes is called psychopharmacology.

Certain patterns of disease with mental manifestations are biologically characteristic of humans. The use of drugs to treat these disease patterns is directed either at alleviating symptoms or at inhibiting or stopping the underlying disease processes. Diseases that have purely psychological causes, but appear in ways that disturb society or distress the individual, are often treated with nonspecific remedies, that either sedate or alert the individual. For all mental illness with specific biological causes, psychopharmacologists seek to develop drugs that change the biological functioning of the individual so that the symptoms of disease either do not occur or have a lesser impact on his or her life and behavior.

The use of drugs in the treatment of psychiatric illness is not a denial of the importance of psychological or social factors in the causation or pattern of a disease. Drug treatment of psychiatric illness is based on the principle that the human nervous system is always a chemical biological system. Some psychiatric treatment systems – for example, psychoanalysis— do not utilize drug treatment. Some mental health experts feel that the use of drugs is only for the control of patients and not their treatment.

Social and Ethical Issues of Drug Use

Conflicting Values in Drug use

The social and economic requirements of modern society may have undergone a radical change in the last few decades, even though the inertia of the existing social character, its desires and its values, will be felt for some time to come. In one major sense, current drug controversies are a reflection of this cultural lag with all of the consequent conflict of wishes and values that result of the consequent conflict of wishes and values that result from the lack of good correspondence between traditional teachings and the view of the world as it is now being perceived by large numbers within society. Modern society is in a state of rapid transition, and this transition is not without its untoward consequences in terms of stability.

Cultural transitions notwithstanding, the dominant social order has strong negative feelings about any nonsanctioned use of drugs that contradicts its existing value system. Can society succeed if individuals are allowed unrestrained self-indulgence? Is is bad to rely on something so much that one cannot exist without it? Is it legitimate to take drugs if one is not sick? Does one have the right to decide for oneself what one needs? Does society have the right to punish someone if he has done no harm to himself or to others? These are difficult questions that do not admit to ready answers. One can guess what the answers would be to the nonsanctioned use of drugs. The traditional ethic dictates harsh responses to conduct that is "self-indulgent" or "abusive of pleasure". But how does one account for the quantities of the drugs being manufactured and consumed today by the general public? It is one thing to talk of the few hundred thousand or so "hard" narcotic users who are principally addicted to the opiates. One might still feel comfortable in disparaging the widespread illicit use of hallucinogenic substances; these are still the "other guys." But the sedatives, tranquilizers, sleeping remedies, stimulants, alcohol, coffee, tea, and tobacco are complications that trap the advocate in some glaring inconsistencies. It may be asked by partisans whether the cosmetic use of stimulants for weight control is any more legitimate than the use of stimulants to "get with it?"; whether the conflict-ridden businessman or the conflict-ridden housewife is any more entitled to relax chemically (alcohol, tranquilizers, sleeping aids, sedatives) than the conflict-ridden adolescent?"; Whether physical pain is an less bearable than mental pain or anguish? Billions of pills and capsules of a nonnarcotic type are manufactured yearly.

Sedatives and tranquilizers account for somewhere around 12 to 20 per cent of all doctors' prescriptions. In addition there are about 150 different sleeping aids that are available for sale without a prescription. The alcoholic beverage industry produces countless millions of gallons of wine and spirits and countless millions of barrels of beer each year. One might conclude that there is a whole drug culture; that the problem is not confined to the young, the poor, the disadvantaged, or even to the criminal; that existing attitudes are at least inconsistent, possibly hypocritical. One always justifies one's own drug use, but one tends to view the other fellow who uses the same drugs as an abuser who is weak and undesirable. It must be recognized that the social consensus in regard to drug use and abuse is limited, conflict ridden, and often glaringly inconsistent. The problem is not one of insufficient facts but one of multiple objectives that at the present moment appear unreconcilable.

35

Rural Poverty in India and Development as a Policy Challenge

Poverty can be overcome, and that the poor can increase their income and production within an appropriate framework. Part of that framework is made up of a flow of resources and local-level institutional development, and there is considerable scope for improvement in both. However, the impact of investment and organization is strictly determined by the nature of the policy environment. While project and programmes can bring some relief to the rural poor, substantial change needs a strong policy commitment. While the poor can overcome poverty, they will not be able to until this becomes a major focus of national policy and action. In the main, this sort of commitment has not been made in the past – at the expense of both the poor and overall development in many areas.

The current state of India is highly contradictory. On the one hand, there is proclamation of a new order; on the other, increasing value is given to sectional and short-term national and group interests. With an overt concern with India's poor goes an equal weight given to concern with economic mechanisms and relations that pay little attention to poverty and foster more inequality. The dangers of this situation are real. The lack of concrete attention being given to change will mean greater economic polarization. Greater polarization among the better – off, and between the better-off and the poor – means instability and a lack of consensus, a lack of legitimacy.

Poverty is far-reaching, and ought to be curtailed. In a period in which resources everywhere appear restricted, this seems not to be an attractive proposition at the practical level. Welfare is everywhere giving way to production as an imperative, just as public expenditure is giving way to private accumulation. Poverty alleviation does not appear to be an idea whose time has come. The objections are great, but they are also misplaced. Poverty alleviation is not necessarily a drain upon accumulation, and it is not primarily a public activity. Poverty alleviation is primarily the activity of the poor themselves, and their progress necessarily involves productive expansion. If this potential for private expansion has not been realized, it is not because of the nature of the poor, it is because of the way in which national economic affairs have been organized. Economic policy has been oriented towards the better-off – not infrequently at the expense of the poor. Given the historic association between wealth and power, the definition of development in terms of the large and the wealthy is hardly surprising.

There is the possibility of associated growth involving both large-scale and small-scale production, the better-off and the poor. The realization of this possibility might result from a new social compact. This social compact is not a commitment to social safety nets and welfare, both of which seem to presuppose that the poor are somehow necessarily out of the growth field. It is a commitment to abolishing artificial and onerous terms of exchange that discriminate against the poor, to investing resources where there are real opportunities for gain, irrespective of whether the economic agents concerned are rich or poor, and to creating the space for the poor to organize to pursue their social and economic interests.

There is a need for a new growth model consistent with new social realities. While the 1980s was a period of clearing away many of the obstacles to development, it was not a period in which there emerged a clear vision of what represented the positive basis for growth, beyond, that is, a general prescription of market-driven operations. The model must pass from admonition to positive prescription to fuel growth by integrating the poor in their rightful place in the production function. It must redefine the position of public expenditure in the development process, and seek to establish market structures which are both equitable and open to the participation of the economically weaker elements of the population. Most of all it must revalue the position and contribution of the poor and small-scale producers in the growth process, particularly in the agricultural sector, but not exclusively agriculture.

This means that the issue is not so much one of less government, but of government, both national and local, finding a new rationale for action, including, *inter alia*, creating conditions that will effectively unleash the productive potential of the rural poor.

Financial flows to the poorest Indians are not likely to undergo a very major expansion, especially through private channels. Development will rely very much on the mobilization of their own resources, and many of these resources are in the hands of the poor, are, indeed, not only the human capital embodied in the poor but also their assets which, while small, individually are cumulatively important in India. The growth model for the 1990s will have to embrace that fact, and build upon it. The paradox of most development models is that they have emphasized the value of what Indians do not have, while devaluing what they have: capital intensity has been promoted in situations of scarcity of capital, at the expense of abundant labour and of low-cost methods of manifold increase of the productivity of assets of which the poor do dispose. In a not very indirect way, the creation of poverty has been subsidized. Poverty alleviation is neither a special topic nor a low-cost substitute for growth. It is neither more nor less "social" than development in general. It is part of the formulation of any sustainable strategy of economic development. In the 1990s it may, and perhaps should, become the dominant issue—not as an alternative to the structural reorganizations of the 1980s, but as a means of filling a growth framework with substance.

Bibliography

Ackoff, R.L., *Redesigning the Future: A Systems Approach to Societal Problems* (John Wiley, 1974).

Adelman, I., *et al. Economic Growth and Social Equality in Developing Countries* (California, Standford University, 1967).

Aggarwal, Y.P., *Education and Human Resource Development* (New Delhi, Commonwealth, 1988).

Amirk Singh, 'New Policy on Education: Two Years Later,' *Economic & Political Weekly,* Special Number, Vol. XXIII, Nos. 45, 46 & 47, pp. 2479-92.

Anand, Mulk Raj, 'A Nation of Illiterates' *The Tribune*. Feb.12, 1991.

Anderson, C.A., 'A Skeptical Note on Education and Mobility' A.H. Halsey & Others (ed)—*Education Economy and Society,* (New York, The Free Press, 1969), pp. 164-182.

Anderson, C.A., 'Access to Higher Education and Economic Development' in *Halsey, A.H. (Ed)-op. cit.* work. pp. 252-268.

Anderson, C.A. and Bowman, M.J., *Education and Economic Development* (Chicago, 1965).

Anon, 'The Pressure of Economic Change' in A.H. Halsey, (Ed), *op,cit.* pp. 22-30.

Anon, 'All-out Bid to Tap Human Resources', *The Economic Times* (Supplement), Dec. 20, 1984, pp.1-3

Asharaya, P., 'Education: Politics and Social Structure', *Economic & Political Weekly,* Vol. XX, No. 42, Oct. 19,1985, pp. 1785-89.

Bantock,G.A. *Education and Values,* (London, Faber & Faber, 1966).

Bauer, R.A. (Ed), *Social Indicators* (Cambridge and London, MIT Press, 1966).

Becker, Garry S., *Human Capital* (Princeton, Princeton University Press, 1964).

Be⌐'er, Garry S., *Human Capital: A Theoretical and Empirical Analysis with Special Reference to Education*, (New York, NBER, 1974).

Ben-Porath, Yoram. 'The Production of Human Capital and the Life Cycle of Earnings', *The Journal of Political Economy*, August, 1967, pp. 352-65.

Benson, Charles S. *Perspectives on the Economics of Education* (Boston, Houghton Mifflin Company, 1963).

Bhalla, G.S. and Bhalla, H.S. 'Human Resource Development for Rural Poor', Paper presented at the U.G.C. *National Seminar*, held at G.K.I.A.S. in Rural Development, Punjabi University Campus, Damdama Sahib).

Bhatia, S.K., 'Challenges in Human Resource Management', *Indian Management*, Vol. 25, No. 8, August 1986, pp. 5-12.

Blaug, Mark (Ed), *Economics of Education-I* (New York, Penguin, 1968).

Blaug, Mark (Ed), *Economics of Education-II* (New York, Penguin, 1969).

Blaug, Mark (Ed), *An Introduction to the Economics of Education* (New York, Penguin, 1970).

Blaug, Mark, 'The Empirical Status of Human Capital Theory: Slightly Jaundiced Survey', *Journal of Economic Literature*, Vol. 14, No. 3, September 1976, pp. 827-55.

Boulding, K., *The Meaning of the Twentieth Century* (London, Allen & Unwin, 1965).

Bowman, M.J., 'Education and Economic Growth *in King, I. (Ed), Education and Income* (Staff working paper No. 402, Washington, World Bank, 1980) pp. 1-71.

Bowman, M.J., 'The Human Investment Revolution in Economic Thought', *Sociology of Education* 39/2 (Spring), pp. 111-37.

Brown, Murraya (Ed), *The Theory and Empirical Analysis of Production* (New York, NBER, 1967).

Brownstein, L., *Education and Development in Rural Kenya* (New York, Praeger, 1972).

Burgess, T., *et al., Manpower and Educational Development in India* (London, Oliver & Boynd).

Byars, L.L. & Rue, L.W., *Human Resource Management* (Illinois, Irwin Homewood).

Chattopadhyay, G., 'Education: The Authority to Learn or the Authority of the Bowl of Hemlock' in *Decision* (IIM, Calcutta), Vol. 16, No. 1, Jan-March, 1989, pp. 22-33.

Cheema, C.S. 'The Challenges of Human Resource Development in Rural Punjab' – Paper presented at U.G.C. *National Seminar* held at G.K.I.A.S. in Rural Development, Punjabi University Campus, (Damdama Sahib).

Clark, Harold F., 'The Return on Educational Investment' in C.S. Benson, (Ed), *op. cit.*, 1963, pp. 24-32.

Coombs, P.H. and Manzoor Ahmed, *Attacking Rural Poverty: Non-Formal Education Can Help* (John Hopkins University Press, 1974).

Coombs, P.H., *The World Crisis in Education: The View from Eighties* (Oxford, OUP, 1985).

Correa, Hector, *The Economics of Human Resources* (Amsterdam, North-Holland, 1963).

Curle, Adam, 'Some Aspects of Educational Planning in Underdeveloped Areas, *Harvard Educational Review*, Vol. 32, No. 3, 1962.

D'Souza, A.A. and De Souza., A. *Population Growth and Human Development* (Delhi, ISI, 1974).

Datta, S., 'Human Resource Development', *Man and Development*, Vol. 8, No. 1, March 1986, pp. 9-17.

Davis, R.G., *Planning Human Resource Development*, (Chicago, 1966).

Davis, Russel G. *Planning Human Resource Development: Education Models and Schemata* (Chicago, CSED, Harvard University, 1966).

Denison, Edward F., 'Education and Growth' in Benson, C.S.(Ed), *op. cit.*, pp. 32-42.

Desai A.R. *Social Background of Indian Nationalism* (Bombay, Popular, 1966).

Deshmukh, C.D. 'Management and Administration: New Trends', *Training Abstracts 17*, New Delhi Training Division, 1972.

Dey, B., 'On Costing Education' in Pandit's *Measurement of Cost Productivity and Efficiency of Education* (New Delhi, NCERT, 1969), pp. 14-26.

Dey, B. 'Training in the Civil Services: Plea for A Holistic Construal', *Indian Journal of Public Administration* Vol. XXIV, No. 4, Oct.-Dec. 1982.

Drucker, Peter F. 'The Educational Revolution' in Halsey and Others (Ed) *op. cit.*, pp. 15-21.

Drucker, Peter, F., *Managing in Turbulent Times* (William Heinemann, 1980).

Dwivedi, R.S., *Management of Human Resources: A Behavioural Approach to Personnel* (New Delhi, Oxford & IBH, 1982).

Farooq, Khan A., 'Development of Human Resources', *The Economic Times*, September 22, 1984.

Gandhi, Rajiv, 'New National Policy on Education', *Inaugural Address* at the Conference of Education Ministers at New Delhi, August 29, 1985.

Gill, K.S., 'Agricultural Development in Punjab' in Johar and Khanna's (Ed), *Studies in Punjab Economy* (Amritsar, GNDU, 1983).

Gore, M.S., 'Literacy: Equaliser of Opportunity', *Democratic World*, March 31, 1991, Vol. XX, No. 13.

Gostkowski. Z. *Towards A System of Human Resources Indicators for Less-Developed Countries* (The Polish Academy of Sciences).

Government of India, *Challenges of Education: A Policy Perspective* (Government of India, Ministry of Education, 1985).

Government of India, *National Policy on Education* (New Delhi, Govt. of India, 1986).

Government of India, *National Policy on Education: Programme of Action* (New Delhi, Govt. of India, 1986.)

Groves, Harold M., 'Education and Economic Growth' in C.S. Benson, (Ed), *op. cit.*, pp. 7-11.

Halsey, A.H. and others (Ed), *Education, Economy and Society* (New York, The Free Press, 1969).

Harbison, F., 'The Prime Movers of Innovations' in Halsey, A.H. & others (Ed) *op. cit.*

Harbison, F. *Human Resources as the Wealth of Nations* (London, OUP, 1973).

Harbison, F. and Myers, C.A. *Education, Manpower and Economic Growth* (New York, 1974).

Havighurst, R.J., 'Education and Social Mobility' in Four Societies' in A.H. Halsey, & others (Ed), *op. cit.*, pp. 105-120.

Heyneman, S.P., *Improving the Quality of Education in Developing Countries* (Washington, World Bank, 1983).

Heyneman, S.P. and White, D.S., *The Quality of Education and Economic Development* (Washington, World Bank, 1986).

Hicks, Norman, *Economic Growth and Human Resources*, World Bank, Staff Paper No. 408, (Washington, World Bank, 1980).

Hilton School of. *Human Resource Development* (Vellore, ISSR, 1989).

Huq, M.S., *Education, Manpower and Development in South and South-East Asia* (Delhi, Sterling, 1975).

Hussain, Majid, *Agricultural Geography* (New Delhi, Inter-India, 1986).

Jagannathan, N., 'Gender Equality in Education', *University News*, Vol. XXIX, No. 5, Feb. 4, 1991, pp. 1-5.

Jamison, D.T. and Laurence, J.L., *Farmer Education and Farm Efficiency* (Baltimore, John Hopkins, 1982).

Jhingan, M.L., *The Economics of Development and Planning* (New Delhi, Vikas, 1975).

Johnson, D. Gale., 'Economics and the Educational System, in C.S. Benson, (Ed), *op. cit.*, pp. 374-80.

Joshi, P.C., 'Role of Culture in Social Transformation and National Integration, *Economic & Political Weekly*, Vol. XXI, No. 28.

Kamat, A.R., *Progress of Education in Rural Maharashtra* (Pune, Gokhale Institute of Politics and Economics, 1968).

Khanna, G., Parkash, S. and Bansal, R.K., *Unit Cost of College Education in Punjab* (Patiala, Punjabi Unversity, 1985) mimeo.

Khullar, K.K., 'Four Decades of Education', *Yojana*, Vol. 33, No. 8, Nov.1-15, 1989, pp. 12-4.

King, T. (Ed)., *Education and Income*, World Bank Staff Working Paper No. 402, (Washington, World Bank, 1980).

Kirpal, P. 'How To Plan Education of The Future', *Yojana*, Vol.33, No.14&15, August, 1989.

Kothari, Commission, *Report of the Education Commission: 1964-66* (Delhi, Government of India, 1970).

Kothari, V.N. and Panchamukhi, P.R., 'Economics of Education: A Trend Report' in *ICSSR's A Survey of Research in Economics* (New Delhi, 1980), pp. 169-238.

Krishnamurthy, H.V. 'Human Resource Development Strategy For 21st Century; *P.U. Management Review*, Vol. 9, Nos. 1 & 2, Jan-Feb. 1986, pp. 79-89.

Kulkarni, V.G. 'Alternatives in Education' *Man & Development* (Vol. VIIII), March 1985, pp. 25-58.

Lipton, *Why Poor Stay Poor: A Study of Urban Bias in World Development* (London, Templesmith, 1977).

MacNamara, R.S. *The Assualt on World Poverty* (Washington, World Bank, 1975).

Mahajan, V.S. 'Whither National Policy: Education', *The Tribune*, Feb. 24, 1991, p. 8.

Majumdar, Tapas. *Investment in Education and social Choice,* (New York, Cambridge University Press, 1983).

Marshall, Alfred. 'Education and Invention' in Benson, C.S. (Ed), *op. cit.*, pp. 82-83.

Mathur, B.L. (Ed)., *Human Resource Development: Strategic Approaches and Experiences,* (Jaipur, Arihant, 1989).

Mathur, R.N., *Population Analysis and Studies* (Allahabad, Chugh).

Megginson, L.C., *Personnel and Human Resource Administration,* 1974.

Mehta, M.M., *Human Resource Development Planning,* (Delhi, Macmillan, 1976).

Mingat, Alan and Tan, Jee-Pang. *Analytical Tools for Sector Work in Education* (Baltimore/London, John Hopkins University Press, 1988).

Mishra, L. 'Literacy: Now or Never' in *Yojana,* Vol. 34, No. 20, Nov. 1-15, 1990, pp. 4-5.

Mishra, S.K. and Puri, V.K., *Development and Planning: Theory and Practice,* (Bombay, Himalaya, 1986).

Moddie, A.D. *Explorations in Management Development* (New Delhi, AIMA, 1976).

Myrdal, G., *Asian Drama* (Penguin, 1963).

Nadler, L., *Developing Human Resources* (Texas, Concepts, 1979).

Nadler, L., *The Handbook of Human Resources Development* (John Willey, 1984).

Nallagounden, A.M. 'Investment in Education in India', *Journal of Human Resource,* Vol. 2, No. 3, Summer 1967, pp. 347-58.

Nandedkar, V.G., 'Human Resource: Both an End and Means', *Yojana,* Vol. 34, Nos. 1 & 2, Jan. 26, 1990, pp. 50-53.

National Council of Educational Research and Training. *The Fourth All India Education Survey* (New Delhi, NCERT, 1982).

Niland, John R., *The Producation of Manpower Specialists: A Volume of Selected Papers* (New York, Cornell University, 1971).

Nurkse, R., *Problems of Capital Formation in Underdeveloped Countries,* (New Delhi, OUP, 1973).

OECD, *Education in OECD Developing Countries: Trends and Perspectives,* (France, OECD, 1974).

Ota, Masao. 'Quantitative Method for the Planning of Human Resource Development', *Research Bulletin of the National Institute for Educational Research,* No. 11, 1972, pp. 25-41.

Panchamukhi, V.R., Leading Issues in Human Resource Development in India' in P.R. Brahmananda, and V.R. Panchamukhi, (Ed), *Development Process of the Indian Economy,* (Bombay, Himalaya, 1987), pp. 1060-1108.

Pandit, H.N., *Measurement of Cost Productivity and efficiency of Education* (Delhi, NCERT, 1969).

Rao, B.S., 'Developing Human Resources in Rural Areas: Some Basic Propositions' in M.K. Rao and P.P. Sharma, (Ed), *Human Resource Development for Rural Development* (Bombay, Himalaya, 1989), pp. 3-6.

Rao, N.K. and Sharma, P.P., *Human Resource Development for Rural Development* (Bombay, Himalaya, 1989).

Rao, T.V., 'Some Thoughts on HRD in Education', *Indian Journal of Training and Development,* July-Sept., 1986, pp. 135-7.

Rao, T.V., *et al., Alternative Approaches and Strategies of Human Resource Development* (Jaipur, Rawat, 1988).

Rao, T.V., 'Planning for Human Resources Development' in Mathur, B.L. (Ed), *op. cit.*

Rao, V.L., 'Human Element in Economic Development' in Rao, M.K. and Sharma, P.P. (Ed), *op. cit.*, pp. 20-27.

Rao, V.K.R.V., *Education and Human Resource Development* (Bombay, Allied, 1960).

Ravishankar,S., *et al., Human Resource Development: In a Changing Environment* (Bombay, Dhruv & Deep, 1988).

Raza, Moonis, *et al., Education and the Future: An Indian Perspective,* A UNESCO Sponsored Study (New Delhi, NIEPA, 1983).

Raza, Moonis (Ed), *Education Planning: A Long Term Perspective* (Delhi, Concept, 1986).

Ruark, Henry C. (Jr.), 'Technology and Education' in *Benson, C.S. (Ed), op. cit.*, pp. 381-92.

Sapra, C.L. and Aggarwal, Y. (Ed), *Education in India: Critical Issues* (New Delhi, National, 1987).

Saxena, J.P., *et al.*, 'Human Resource Development Strategy', for India's Seventh Plan', *The Economic Times,* July 25, 1984.

Schelsky, H., 'Technical Change and Educational Consequences', in Halsey (Ed), *op. cit.*, pp.31-6.

Pandit, H.N., 'A Study in Unit Costs at School Stage in India: A Design of the Research Project (Pandit, H.N. (Ed), 1969, *op. cit.*, pp.3-13.

Panigrahi, D., 'Human Resource Development in Business Administration', B.L. Mathur, (Ed), *op. cit.*

Parkash, Shri., *Educational System of India: An Econometric Study* (Delhi, Concept, 1978).

Parminder Kaur, and Singh, Bhawdeep, 'Human Resource Development in Rural Punjab' in *U.G.C. National Seminar* held at G.K.I.A.S. in Rural Development (Punjabi University Campus), Demdama Sahib, 22-23 Feb., 1991.

Patel, S.J., 'Educational Miracle in The Third World', *Economic and Political Weekly*, Vol. XX, No. 31, August 3, 1985, pp. 1312-17.

Patil, V.T. and Patil, B.C., *Problems in Indian Education* (New Delhi, Oxford & IBH, 1982).

Perlman, R., *The Economics of Education: Conceptual Problems and Policy Issues* (McGraw Hill, 1973).

Paillai, S.S., 'Education System and Social Structure', *Education India*, Vol.9, March, 1973.

Planning Commission, *The Sixth Five Year Plan: 1980-85* (New Delhi, Government of India).

Planning Commission., *The First Five Year Plan: 1951-56* (Delhi, Government of India).

Psachorapoulos, G., *Earnings and Education in OECD Countries* (Paris, OECD, 1973).

Psachorapoulos, G., 'Education and Development —A Review', *Pigmy Economic Review*, Monthly Economic Journal of the Syndicate Bank, Oct. 88, Vol. 34, No. 3.

Punit, A.E., *Social System in Rural India* (New Delhi, Sterling, 1978).

Radhakrishnan, S., *The Creative Life*.

Index

R

S

T